Big Data Analytics 2017

AA009121

Topical Conference at the 2017 AIChE Spring Meeting and
13th Global Congress on Process Safety

San Antonio, Texas, USA
26 – 30 March 2017

ISBN: 978-1-5108-4132-1

Printed from e-media with permission by:

Curran Associates, Inc.
57 Morehouse Lane
Red Hook, NY 12571

Printed by Curran Associates, Inc. (2017)

AIChE
120 Wall Street, FL 23
New York, NY 10005-4020

Phone: (800) 242-4363
Fax: (203) 775-5177

www.aiche.org

Additional copies of this publication are available from:

Curran Associates, Inc.
57 Morehouse Lane
Red Hook, NY 12571 USA
Phone: 845-758-0400
Fax: 845-758-2633
Email: curran@proceedings.com
Web: www.proceedings.com

TABLE OF CONTENTS

482030 AGILE Award Keynote Address

Abstract:

Factors in Selecting AGILE Honoree

1. Overall sustained excellent industry leadership in businesses that employ chemical engineers

2. Recognized for efforts in innovation and creating strategies that ensure success

3. Value of the keynote to the meeting attendees and the members of the Institute Schedule

Event	Time	Location
Continental Breakfast	7:00AM - 8:00AM	Henry Gonzalez Convention Center, Lila Cockrell Theatre
AGILE Keynote Address	8:00AM - 9:15AM	Henry Gonzalez Convention Center, Lila Cockrell Theatre

The 2017 AIChE Spring Meeting and 13th GCPS AGILE Keynote Address will be presented by AGILE award recipient John Y. Televantos, Partner, Arsenal Capital Partners.

John Televantos is a Partner at Arsenal Capital Partners, a middle market private equity firm focused on the specialty industrial, healthcare and specialty financial service sectors. Before joining Arsenal in 2006, Dr. Televantos was President of Aqualon Company, owned by Hercules, Inc.; President and Founder of Helios Chemical Company; and Chief Executive Officer for Foamex International.

Dr. Televantos began his career with Union Carbide in 1977 in silicones research. There he developed two key processes for the manufacture of Methyl Chloride and Siloxane polymers that were commercialized. These processes not only improved efficiency and reduced cost but also dramatically reduced waste. He moved through a variety of management positions there and later at ARCO Chemical, where he became vice president of strategy and development.

At ARCO John co-invented and led a team that developed and commercialized a breakthrough technology in polyol manufacture (a key raw material for polyurethanes) eliminating a whole process step along with waste and emissions, while at the same time opening the door to the creation of new higher-performance polyurethanes. The technology has since been licensed by most of the major global polyol manufacturers with significant waste reduction worldwide.

At Arsenal Capital, Dr. Televantos serves as co-head of the specialty industrial practice. During his tenure, he has been part of several acquisitions and successful exits in the chemicals and materials sectors. In addition to serving on the boards of several Arsenal-owned companies, he is lead director on the board of Balchem Corp.

480703 Big Data Analytics Skills - a Revolution Lacking in Revolutionaries

Abstract:

Since the advent of the common use of the term "Big Data[i]" in 2010 to describe large, *unruly* sets of data, and the artful analysis thereof, industry has felt besieged by opportunity. The chemical and refining industry has been faced with "Big Data" long before it was cachet. Yet within industry there is still a struggle to get hearts and minds around what the Big Data Journey means to an organization and how to take advantage of the data opportunities that lie within. The added complication to the chemical industry is that it must add "contextualization" to the data analysis because it's not enough just to know there is a signal, we need to know what the signal means[ii]. Yet during this data revolution there is an underlying challenge that Academia, Industry and Governmental Agencies must tackle together to ensure the US Chemical industry retains its competitive position in the world. Together we need to provide education for the higher level skills and abilities required across roles in manufacturing to assure that no one gets left behind in the new Big Data age. Indeed, the next great technological "step" made may not be in engineering design or IoT connectedness (reference Industry 4.0), it may be in designing and creating a better human capability in "data acuity" out of our university system; operators, engineers and scientists alike who are better prepared for the data age than their predecessors. This talk will take a look at what Industry requires in skills sets and perspective to be "data ready", and what Academia and Government might do to improve the output of data-ready skills.

---

[i] Big Data, What is it? White, Chemical Engineering Progress, AIChE March 2016.

[ii] Big Data, Getting Started on the Journey. Colegrove, Seasholtz, Khare, Chemical Engineering Progress, AIChE March 2016.

Abstract

Achieving the next level of asset integrity with predictive maintenance

Exceptional organizations manage to deliver superior operational performance without compromising on safety and integrity. On the contrary, high performing players are at the forefront of equipment reliability for critical equipment and safety performance.

In this context, introducing Predictive Maintenance can dramatically help achieve the next level of asset integrity by leveraging data - structured and unstructured, machine and non-machine based – in order to generate insights with advanced analytics methods that would not be possible with conventional techniques
Implementing Predictive Maintenance generates three benefits that reinforce one another:
Predict when equipment will fail and help reduce the safety and integrity threats
Avoid failures and positively contribute to reduce the scale of the value at risk
Enhance specific asset knowledge and learn more about the behavior of critical equipment

In a recent client project for an upstream oil and gas facility, the main gas compressors have been identified as **critical equipment**: they account for many failures and a large portion of production losses. A Weibull analysis showed that out of several known failure modes, some show a random failure pattern, leading to sudden breakdowns releasing high energy and thus representing a safety issue. As per the random failure behavior, classical time- or usage-based maintenance strategies are not fruitful.

Combining 100's of gigabytes of loss data, maintenance work order data and sensor data from 1,200 sensors around the equipment, a data driven model has been created showing that out of the 1,200 sensor sources, 43 explanatory variables predict the time to failure with ~80% accuracy. The model has been tweaked further in order to reduce the number of false negatives while accepting by this a certain number of false positives (i.e. false alarms) due to safety reasons.

As a result of this model, an early warning of an imminent failure means repairs can be planned, spares can be gathered and the equipment can now be shut down in a safe and orderly manner. Overall, this represents an average reduction from 14 to 4 days of downtime per breakdown event.

Outline

- An organization's ability to maintain asset integrity and anticipate equipment failure positively affects its safety performance
- Predictive Maintenance makes it possible to increase asset integrity and improve operational performance using the millions of operational data points that are being collected
- Example of a success story: Early warning of an imminent failure for a gas compressor on a rig

McKinsey & Company 2

There is a strong correlation between equipment reliability and safety performance

1993 - 2003 (base 100)

EXAMPLE FROM ALUMINUM SMELTING

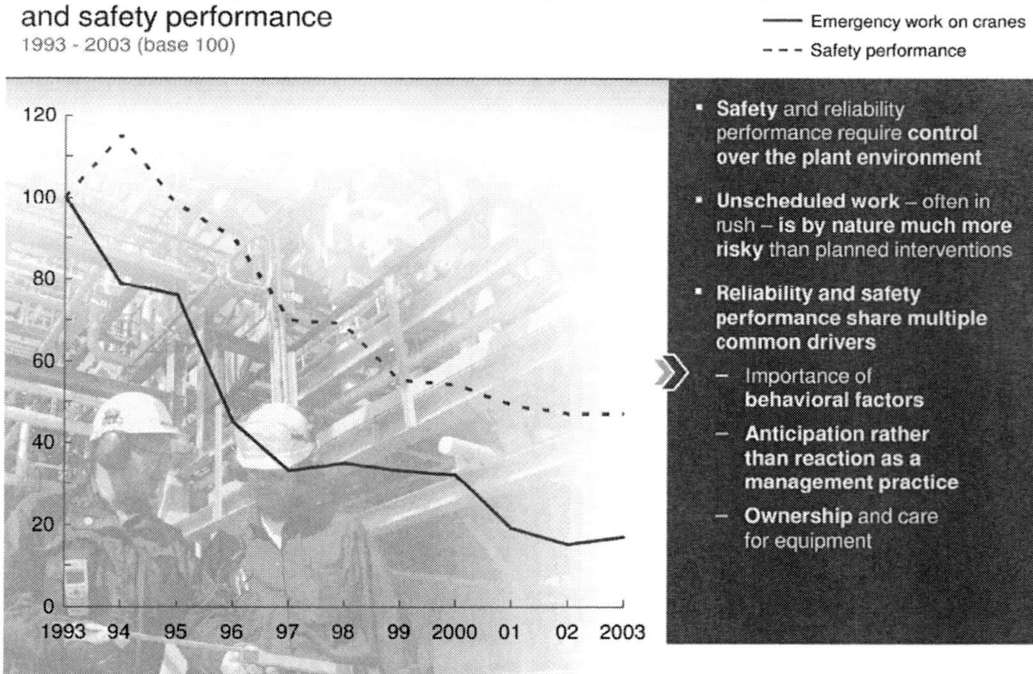

SOURCE: "Practices to Reliability Performance" team

- **Safety** and reliability performance require **control over the plant environment**
- **Unscheduled work** – often in rush – **is by nature much more risky** than planned interventions
- **Reliability and safety performance share multiple common drivers**
  - Importance of **behavioral factors**
  - **Anticipation rather than reaction as a management practice**
  - **Ownership** and care for equipment

McKinsey & Company 4

Industry collects Petabytes of data which can be used for Predictive Maintenance to improve asset integrity and reliability

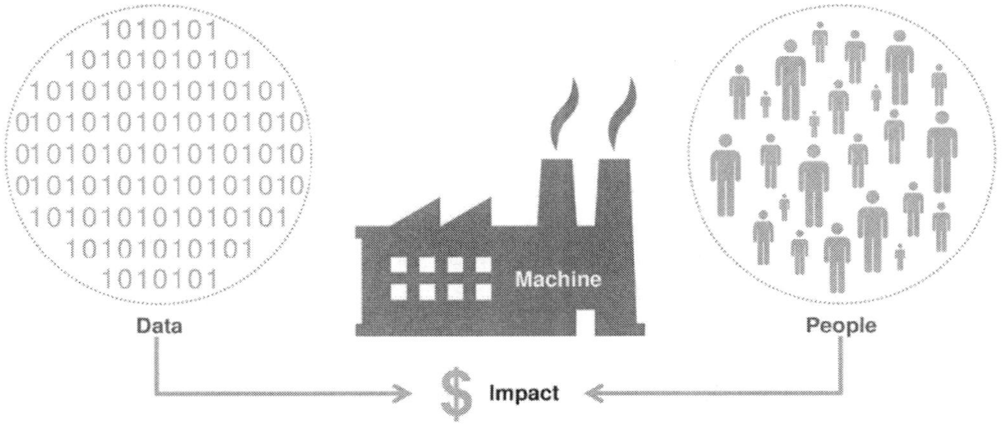

SOURCE: McKinsey

McKinsey & Company 5

Getting from data to impact requires a combination of many different skills

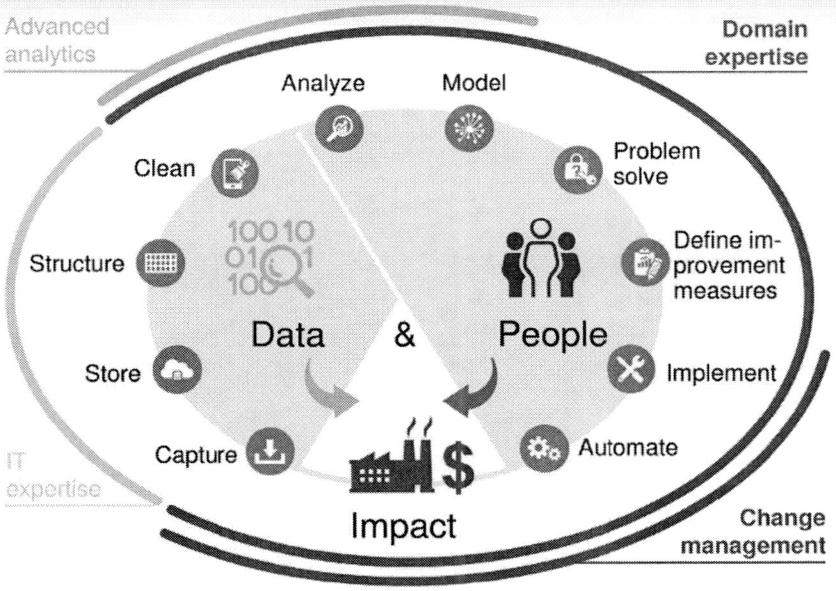

SOURCE: McKinsey

McKinsey & Company 6

Advanced Analytics generates previously untapped insights in Predictive Maintenance from troves of historical data

Predictive Maintenance
is the use of data (structured and unstructured, machine and non-machine based) to generate insights with advanced analytical methods that would not be possible with conventional techniques

Predict

Failure forecasting: Model future asset behavior to anticipate needs *(e.g., HR, spare parts inventory, capital replacement)*
Incipient failure detection: Continuous data acquisition & analysis to detect or predict failure *(e.g., early warnings to allow planned intervention)*

Avoid

Avoid failures
Identification of undesired patterns in data as primary causes of failure (e.g., preventive actions in specific locations of roads with poor conditions to prevent truck failures)

Learn

Leverage specific asset knowledge
Access reliability data from a global equipment database comprised of hundreds of thousands of components across the globe *(e.g. Failure rates, failure mode distributions, causes)* to inform maintenance strategies

SOURCE: McKinsey

McKinsey & Company 7

The main benefits of implementing predictive maintenance include availability increase and cost reduction

Predictive maintenance aims to ...

... detect early signs of problem **for timely correction** at **minimal costs**

... prioritize and optimize maintenance resources using real-time data

... increase unit availability due to **shorter downtimes**

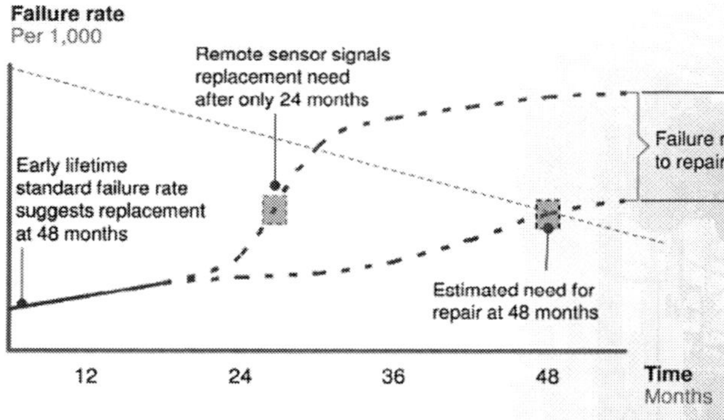

Typical impact of PdM

Machinery breakdowns	-50%
Spare part inventory	-20%
Total machine downtime	-50%
Overtime expenses	-20%
Machine life	+20 - 40%
Overall productivity	**+20 - 30%**

SOURCE: ARC Advisory Group; Department of Defense; expert interviews; GE; Schaeffler; VibroSystM; WiHART Systems

McKinsey & Company 8

Predictive maintenance is proven to add value in many industries

What was Predictive Maintenance used for

✓ Focus of the engagement
Detailed in following pages

Industry	Reduce operating costs	Reduce maintenance costs	Increase machine availability	Impact
1 Mining			✓	▪ 25% downtime reduction
2 Mining *(Copper mine)*	✓	✓		▪ Savings of ~12% in operational expenditures
3 Oil and gas *(Rig operator)*		✓		▪ Reduced maintenance spend by 18 to 27%
4 Oil and gas			✓	▪ ~70% reduction in downtime on critical component
5 Railroads			✓	▪ Reduced failure cost by 20% to 30%
6 Heavy-Motors *(Cruise operator)*	✓	✓		▪ Reduced operating cost by 7%
7 Semiconductor	✓			▪ Increased failure prediction to 87%

We have more than 20 use cases. The remaining cases cannot be shared due to confidentiality.

SOURCE: McKinsey Asset Productivity Service Line

Summary of the case

Applying advanced analytics in off-shore oil and gas gives early failure warning can reduce production losses due to gas compressor by ~70%

Situation

Sector and site
- Offshore oil and gas operator
- Rig maintenance

Situation
- Gas compressors are **critical equipment**: they account for many failures and a large portion of production losses
- Loss data, maintenance work order data and sensor data is collected

Question
- To what extent can we reduce maintenance spend and improve reliability by predicting when a failure will happen?

Typical failure mode: inlet strainer

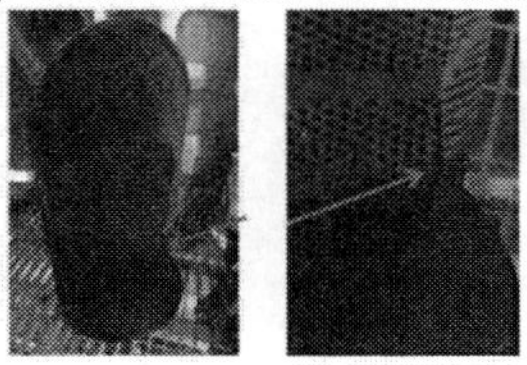

Key Activities

Data collection from several sources
- 120 gigabytes of data
  - ~1,200 sensors
  - 10 years of work orders and deferment data

Identify failures and construct target variable
- Identify if downtime is caused by compressor

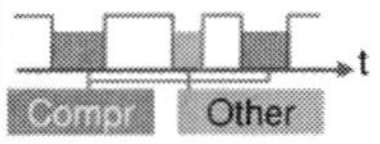

Create a model to predict time to failure
- For any point in time we predict the **time to the next failure**

Inputs 1,200 sensors → **Prediction Model** → **Time to next failure**

- **Inputs** read from 1,200 sensors
- **Regression Model** predicts the time to failure with ~80% accuracy

Impact

- Downtime due to gas compressor is expected to reduce by **~70%** by planning intervention when a failure is imminent
  - **Prediction of time to failure:** if intervened at predicted TTF < 5 days, you would act between 9 and 6 days before the failure actually will happen

Predicted TTF for different failures, Days

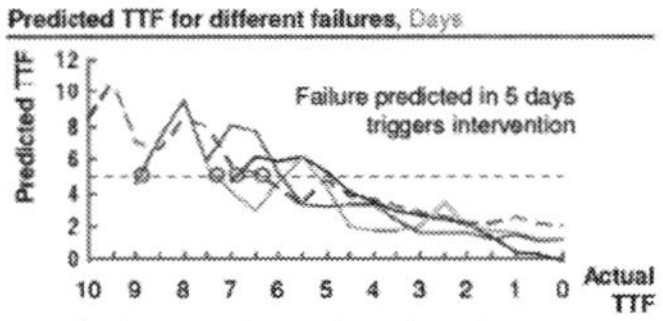

  - **Early warning of an imminent failure** means repairs can be planned. Reduction is on average from 14 to 4 days of downtime

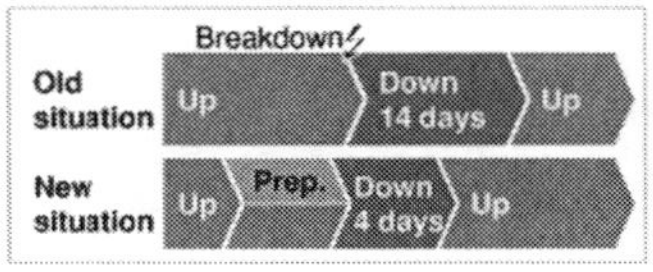

1 TTF = Time To Failure, each of the lines in the graph is one failure case

SOURCE: McKinsey

Data collection was based on 120 GB of data – including more than 1200 sensors and 10 years of work orders

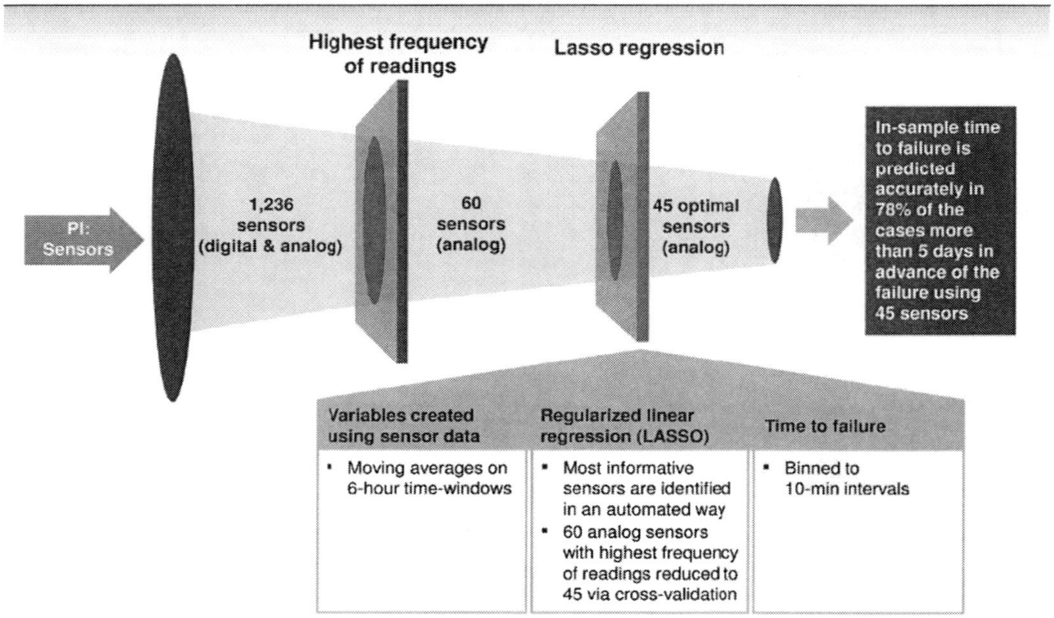

In order to predict downtimes, data preparation as well as a clear target variable were keys to success

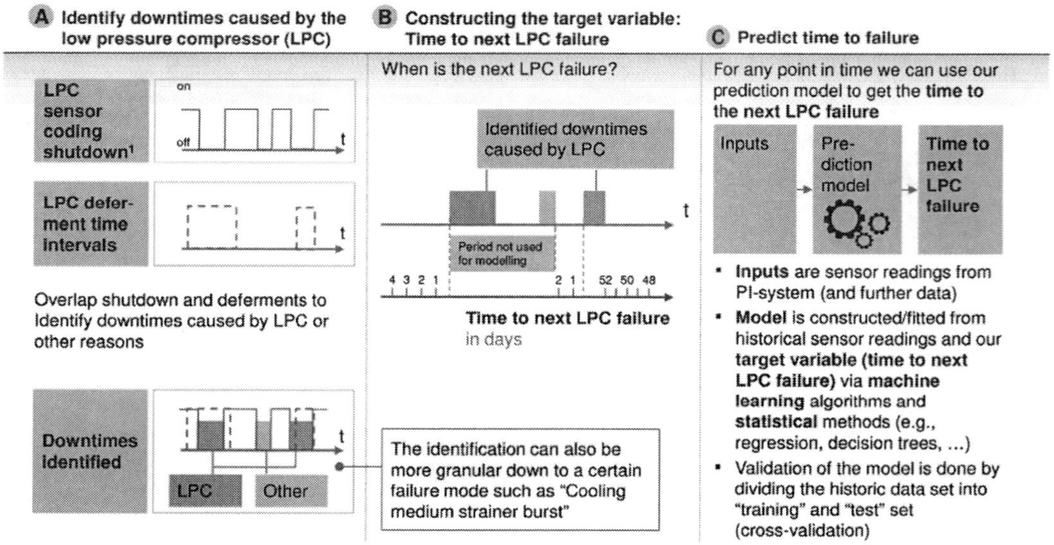

FROM ANALYSIS TO VALUE CREATION

Top down value assessment for predictive maintenance

PRELIMINARY

Digital asset management

- ↑ **Availability 2-5%**
  - ↓ **Planned downtime 10-20%**
    - Decrease in work order frequency
      - Work order cadence to equipment needs instead of OEM recommendation
    - Decrease in work order duration
      - Word order instructions pushed to mobile devices
      - Improved preparation to increase employee wrench time
      - Fully digital and mobile enable maintenance process and cycle
  - ↓ **Unplanned 30-50%**
    - Actively performing routine work orders based on predictive analytics
    - Monitoring conditions of equipment to accurately predict failures
- ↓ **Maintenance cost, 5-25%**
  - ↓ **Operating maintenance costs**
    - Cheaper, routine work orders (predictive > reactive)
    - Improve part and tool inventory strategy with improved database
  - ↓ **Sustaining capex costs**
    - Longer equipment life through planning component replacement extension
    - Adjust to a smaller fleet size

McKinsey & Company 14

I The predictive model combines sensors in a way that cannot be replicated manually

Most important sensors for prediction

10 most significant out of a total of 45 sensors

Rank	Sensor	Unit of measure	Relative contribution Abs. value of coefficient
1	SEAL VENT	BARG	4.2
2	SEAL GAS VENT DP	BAR	2.3
3	FUEL GAS PRESSURE	BARG	1.9
4	COMPRESSOR FLOW DP	MBAR	1.8
5	INLET FILTER DP	MBAR	1.6
6	SEAL GAS SUP PRESS	BARG	1.3
7	ENG BRG2/3 DRAIN	DEG C	1.3
8	FUEL SYS AIR SUPPLY	BARG	1.2
9	FUEL GAS FLOW DP	MBAR	1.1
10	PRI NDE SEAL VENT	BARG	1.1

- The statistical model has identified a total of **45 sensors** to be significant when predicting compressor failures
- The contribution of the sensors is evenly distributed – there is **no one sensor that can replace the predictive model**, which combines a total of 45 sensors
- The absolute value of a coefficient can only be extracted in a linear regression case, and that other metrics should be used for other models e.g. variable importance in random forests

SOURCE: Team analysis, Sensor data

McKinsey & Company 15

II We are able to predict breakdowns at least 5 days before failure

Predicted time until failure. 12-hour average of prediction

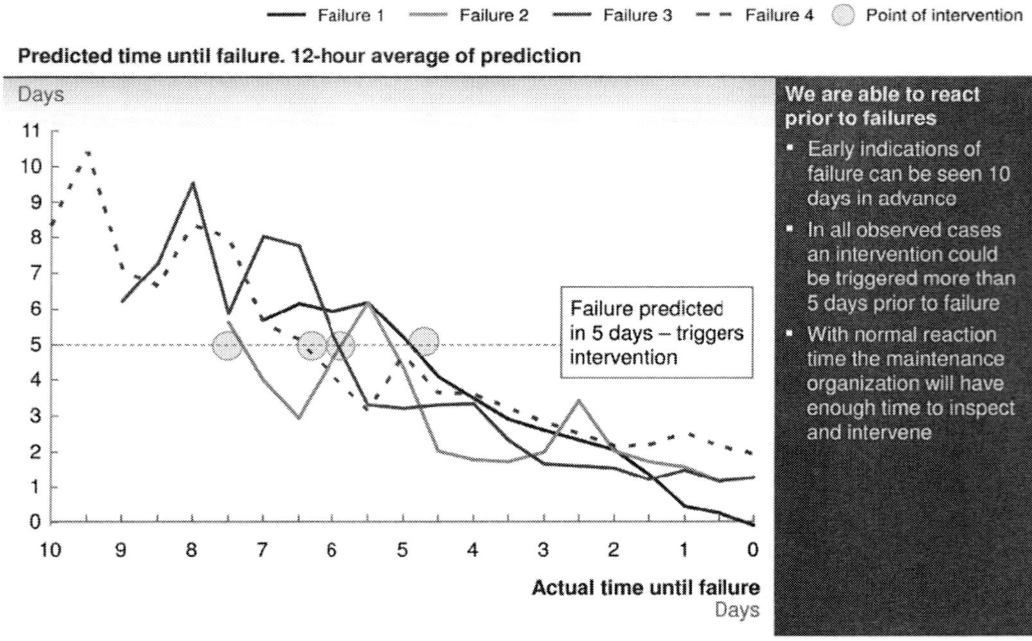

We are able to react prior to failures

- Early indications of failure can be seen 10 days in advance
- In all observed cases an intervention could be triggered more than 5 days prior to failure
- With normal reaction time the maintenance organization will have enough time to inspect and intervene

SOURCE: Team analysis, client data

McKinsey & Company 16

479971 Big Data Vs. Little's Law: A Systems-Based Framework Changing the Risk-Reward of Large Scale Facilities

Abstract:

The current rush to adopt Big Data ignores the fact that *the data is not the system*. Ignoring system dynamics in the interpretation of thedata often produces the opposite of the intended result. Fortunately, Little's Law ($L = \lambda W$) provides a framework to properly interpret the dynamic nature of data and diagnose the true underlying systemic problems. Little's Law is especially useful in identifying and eliminating *hidden factories* in the system that erase profits, increase systemic risk, and erode morale and culture. Dr. Carrier will explore examples from high-risk environments that have saved hundreds of millions of dollars per year while improving safety. In addition, the amount of data required to make and implement informed decisions is dramatically decreased through the use of Little's Law. The approach will be demonstrated with several examples in Oil and Gas, Refining, and Fine Chemicals worked on by the author. The methodology will also be demonstrated for use in maintenance management and the radical improvement of a mid-sized refinery's MoC (management of change process). Finally, the role of culture in actively the dynamic response of our systems will be discussed.

479057 Big Data Doesn't Necessarily Mean Big Intelligence

Abstract:

Big data is the IT phrase of the moment in almost every industry. The concept of collecting huge amounts of data to learn, manage and drive your business forwards is attractive to all managers, especially as without it you may feel that you're being left behind. But without knowing how to collect data effectively, knowing what data has value, and how to effectively and efficiently use the volume of information, Big Data can be a hindrance, not a help. In Process Safety, we rely on data coming from the field in many instances. Collecting this data can be time consuming and costly, so making the most of the data you have is paramount, as is ensuring the data you collect is seen as a valuable asset by everyone in your organisation. But turning that data into intelligence to guide you on your path to Process Excellence is what it's really all about. Data is as valuable an asset as any other. Look after it. Nutrue it. Value it. And make sure you use it. In this paper we examine the core principles of making PSM data work for you to give you the best usable, actionable intelligence possible.

INGENERO

Excellence Through Insight

Industrial IOT with Reliability and Predictive Models to Improve Furnace Reliability and Performance

2017 AIChE Spring Meeting in San Antonio, TX

Session: Industrial Internet of Things I

Sameer Thorat & Jim Brigman

(Ingenero Inc. Houston)

Opportunities "Big Data Analytics" and "IIoT"

Improve machine/system efficiencies by application of solutions arising from internet based innovations, achieve improvements that are challenging to accomplish using the traditional approaches

- The key towards the success of this new multi-dimensional approach principally depends on
  - ✓ Use of advanced analytical tools
  - ✓ In-depth specific domain expertise
  - ✓ Information – usage and automation
  - ✓ Predictive capabilities and Accuracy
  - ✓ Usage towards intelligent decision making
  - ✓ Continuous tracking and optimization

Leverage the power of both historic and real-time data with industry specific advanced analytics

INGENERO

IPOG™ Methodology (DMAIC approach)

Big Data Analytics is combined with IIoT to effectively implement the DMAIC approach to glean value and sustain

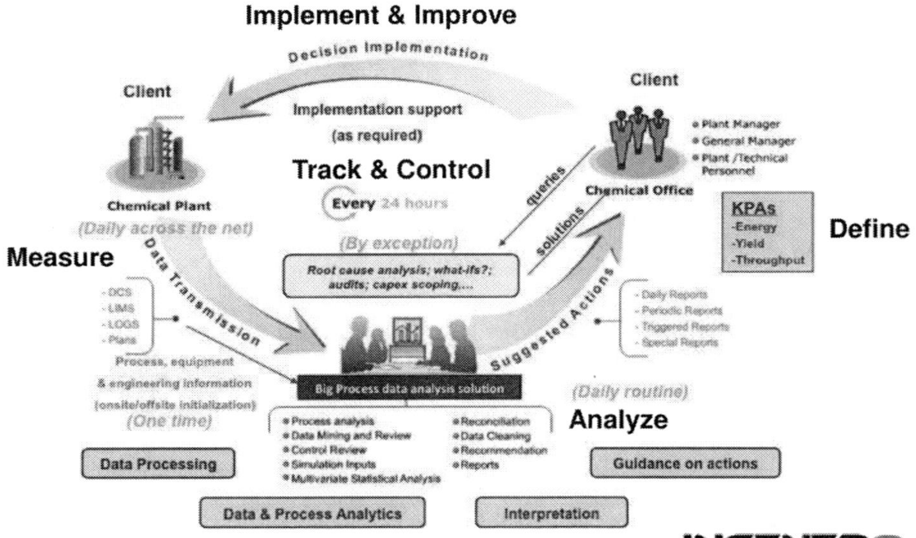

INGENERO

Furnace Performance Improvement Impact

Furnace is core part of ethylene plant operation and any improvement in it's Performance and Reliability significantly helps in increasing production and reducing maintenance and operating costs

- Ethylene production increase
  - ✓ Furnace throughput increase
  - ✓ Furnace yield improvement
  - ✓ Furnace runlength and reliability improvement (increased uptime/availability)
- Cost reduction
  - ✓ Energy utilization efficiency and recovery efficiency
  - ✓ Decoke costs
  - ✓ Effective cost based on tube replacement frequency
  - ✓ Effective cost based on unforeseen shutdown frequency
  - ✓ Maintenance costs

INGENERO

Typical Cracking Furnace Performance Considerations

Ethylene plants tend to have common furnace section issues, which limit performance, of which lower than expected runlength is the most prevalent

- Furnace runlength and capacity gets typically limited by
  - ✓ Coil Pressure Ratio (CPR/CIP)
  - ✓ Radiant Coil Tube Metal Temperatures (TMT's)
  - ✓ Transfer Line Exchanger (TLE) pressure drop or temperature
  - ✓ Convection Bank constraints like tube TMT's and SCR bed temperature
- Few other factors that impact furnace operation
  - ✓ Raw material limitations
  - ✓ Down stream unit requirements/constraints
  - ✓ Overall economics
  - ✓ Maintenance and Decoke scheduling

INGENERO

Addressing The Need Through Furnace Analytics

Addressing the issues requires close monitoring of furnace as well as its downstream parameters, an early warning mechanism and continuous multi-dimensional data analytics

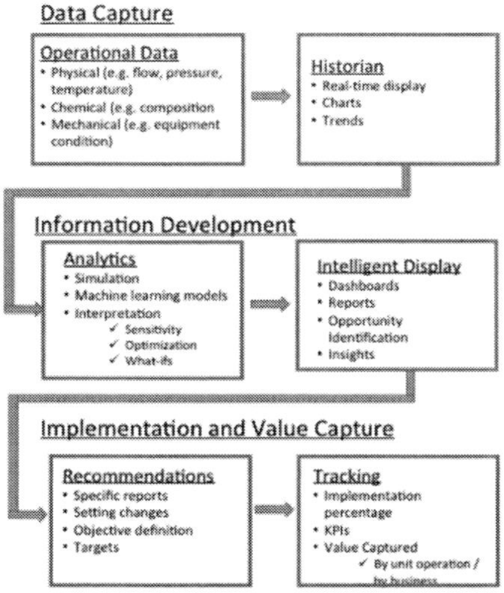

The complete analytics process starts with collection of appropriate data, development of information through analytics (the core), proper visualization and prescriptions, and a tracking and feedback mechanism to convert the learning to wisdom

Data → Information → Knowledge → Wisdom

INGENERO

Methodology ("Big Data Approach")

Implementation Of An Effective Analytics Program

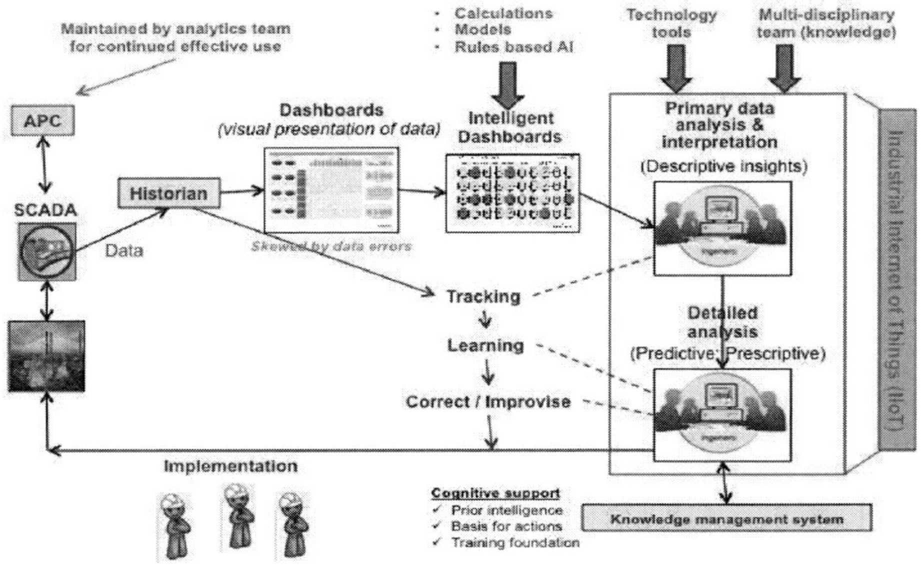

INGENERO

Confidential & Proprietary 7

Models At The Core Of Ethylene Plant Analytics

Different types of models form an essential part of the process, equipment and compliance analytics

- Fundamental first principle Yield models (reaction kinetics, mass & heat transfer, thermodynamics)
  - ✓ COILSIM1D
- Steady state process simulation & optimization
  - ✓ Aspen suite, Hysys, PRO II, FRNC, HTFS, HTRI
- Computational Fluid Dynamics (CFD)
  - ✓ Fluent, OpenFOAM
- Data driven models
  - ✓ Statistical, neural net, pattern recognition, Weibull analysis, PCA, PLS, time series
- LP models
  - ✓ Integrated supply chain model (supply, production, inventory, distribution), strategic planning models, scheduling models, Optimal feed distribution
- Dynamic models
  - ✓ Operator training, model predictive control

INGENERO

Confidential & Proprietary 8

Appropriate Model Technology Utilization

Use of appropriate models and technology is key to help the plant to maximize profit/production and extend furnace run lengths

- Reaction kinetics model to compute yield and coking rate as a function of key parameters like Feed type and composition, Steam to Hydrocarbon ratio, Coil Outlet Temperature, Coil Inlet Pressure, Coil Outlet Pressure
  - ✓ Parallel data driven machine learning models allow early warning indications and quick primary adjustments, which can be cross verified by kinetic models
- Machine Learning data driven models to predict and adjust run lengths to plan decoking cycle
- Convection section modeling (e.g. FRNC 5, HTRI Fired Heaters) to achieve maximum throughput / cracking safely and reliably
- Fire box modeling (geometry / heat flux) to control rate of fouling at the individual coil level
- CFD to address flue gas channeling, flame impingement and hot spots on tubes and flow pattern in Transfer Line Exchanger inlet cone
- LP Model to maximize production options appropriately

INGENERO

Furnace Yield Prediction Models

Predictions provide insights on impact of feed, cracking severity and other production parameters and thereby allow appropriate adjustments to be made to optimize performance and production

- Furnace Models help identify differential cracking among coils

Furnace Coilwise Imbalance - Impact on Conversion, Yields												
	Case-1: Constant severity for differential COT's						Case-2: Differential severity at constant COT's					
	Coil AB	Coil BA	Coil BC	Coil CB	Coil CD	Coil DC	Coil AB	Coil BA	Coil BC	Coil CB	Coil CD	Coil DC
HC Feed Rate	XX	XX	XX	XX	XX	XX	XX	XX	XX	XX	XX	XX
SHC	0.50	0.45	0.45	0.45	0.45	0.45	0.50	0.45	0.45	0.45	0.45	0.45
CIP	39.5	42.0	43.1	41.3	42.8	39.6	39.5	42.0	43.1	41.3	42.8	39.6
COP	16.50		14.61		15.37		16.50		14.61		15.37	
Severity Plant Tag	66.6		-	-	-	-	66.6		-	-	-	-
Severity Calculated	66.6	66.6	66.6	66.6	66.6	66.6	67.8	67.3	65.9	66.3	66.3	67.2
Actual COT(Plant Tag)	1561	1561	1561	1561	1561	1561	1561	1561	1561	1561	1561	1561
Calculated COT(Coilsim)	1556	1558	1564	1563	1562	1559	1561	1561	1561	1561	1561	1561
Heat Absorbed	-	-	-	-	-	-	-	-	-	-	-	-
Effluent (Wt%)												
H2	3.91	3.90	3.89	3.91	3.89	3.90	3.97	3.93	3.86	3.89	3.88	3.93
CH4	6.37	6.50	6.46	6.38	6.49	6.42	6.64	6.64	6.31	6.31	6.44	6.53
C2H4	▲51.17	▼50.99	▬51.14	▲51.25	▼51.02	▬51.15	▲51.88	▬51.37	▼50.72	▼51.05	▼50.87	▬51.48
C2H6	32.63	32.60	32.59	32.59	32.65	32.59	31.41	31.96	33.28	32.92	32.89	32.06
C3H6	1.32	1.36	1.35	1.33	1.36	1.34	1.34	1.37	1.34	1.33	1.36	1.34
C3H8	0.24	0.25	0.25	0.25	0.25	0.24	0.24	0.24	0.25	0.25	0.25	0.24

INGENERO

Furnace Runlength Prediction – Using Kinetics Model

Furnace Reaction Kinetics model is used to predict furnace runlength by estimating the coking rate based on the key operating parameters

- An algorithm coupled with Furnace model fine tuned for actual performance is used to predict coke thickness based on feed composition and rates, cracking severity, steam ratio etc.
- Model estimates key furnace operating parameters Coil inlet and outlet pressures and TMT's which primarily determines furnace EOR criteria

Coil Outlet Pressure (Predictions)				
Actual COP		Predicted COPs		
DOL	5	15	30	40
TLE-X	11.8	12.0	12.3	12.5
TLE-Y	11.5	12.2	12.7	13.1
TLE-Z	12.0	12.0	11.9	11.9

Coil Inlet Pressure (Predictions)				
Actual CIPs		Predicted CIPs		
DOL	10	20	30	40
Coil-A	35.7	39.3	40.7	42.3
Coil-B	35.0	37.8	38.9	40.2
Coil-C	37.3	41.6	44.6	48.5
Coil-D	35.7	37.2	37.8	38.4

TMT, Coke and Yield Predictions			
Coil No.	Max TMT °F EOR	Avg Coke Thickness	Ethylene Yield WT%
Coil-A	1914	0.0092	49.74
Coil-B	1905	0.0087	49.87
Coil-C	2009	0.0146	49.53
Coil-D	1902	0.0070	50.06

INGENERO

Furnace Runlength Prediction – Application

Model prediction results helps in detecting early anomalies, consider process changes to optimize performance, planning and scheduling

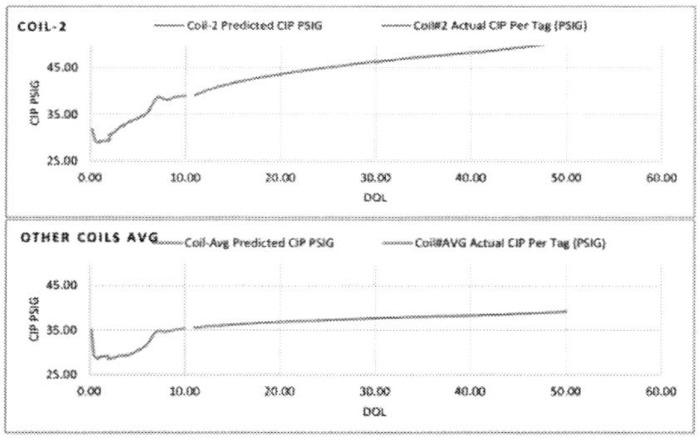

With current operating scenario

→ Coil 2 is expected to have a higher coking rate as compared to other coils

→ Coil 2 CIP will cross 50 psig after ~51 days of operation

→ Coil 2 TMT is expected to cross 2000°F limit after 47 days of operation

Path forward:

→ Analyze reasons for higher coking rate in Coil 2

→ Consider applicable operation changes to minimize the impact

→ Deliberate feed re-distribution and severity changes while evaluating economics

→ Appropriate changes in decoke and maintenance schedule

INGENERO

Convection Section Models

Convection models aids in thermal efficiency improvement, evaluating individual bank performances, determining parameter settings for maximum safe production, identifying fouling and flue gas channeling

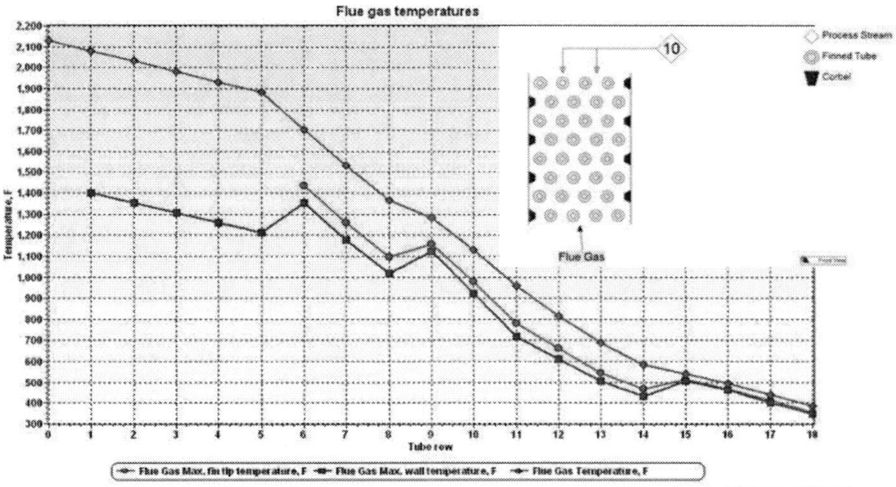

INGENERO

Statistical Models – PCA For Outlier detection

Statistical models based on PCA (Principal Component Analysis) are developed for furnaces, used for identifying the deviations in normal operation/performance and aid improvement

- Key features and benefits of PCA models
  - ✓ Multivariate correlations between several variables captured easily
  - ✓ Compares current furnace run with the past performance
  - ✓ Possibility of missing out on outliers/abnormal operation is very rare
  - ✓ Which variable caused the sample to be an outlier?
    - ❖ Q contribution plot gives that insight
  - ✓ Model is linked to Process Historian and updates automatically
  - ✓ Filtering out outliers is easy and less time consuming
  - ✓ Aids in early detection of faults, improves reliability

INGENERO

Case Study: Early Detection Of Measurement Drift -1

Abnormal deviation in furnace operating parameters was accurately detected and subsequent actions helped in reducing differential coking and thereby extending furnace run length using PCA analysis

PCA - Contribution Plots

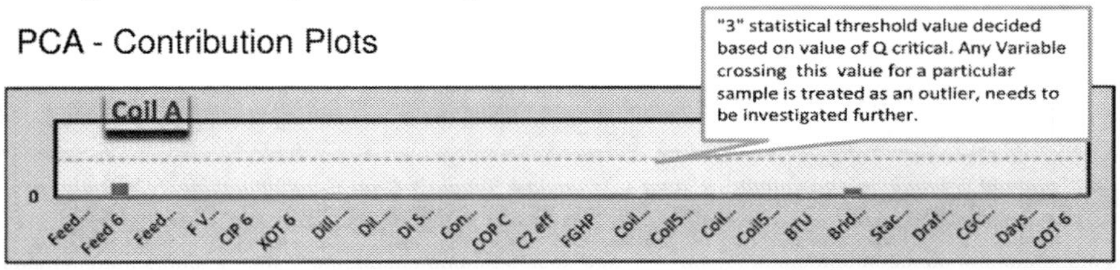

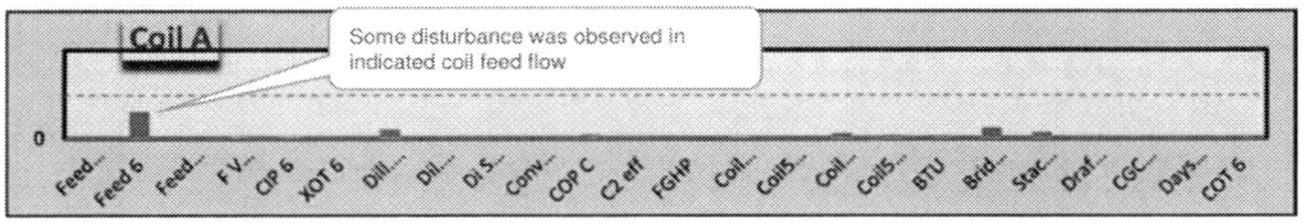

PCA approach saves the effort of looking at multiple trends for multiple variables and judging faults or abnormal operation based on "mental" models

INGENERO

Case Study: Early Detection Of Measurement Drift -2

Model predictions ~ 25th day: Differential coking observed

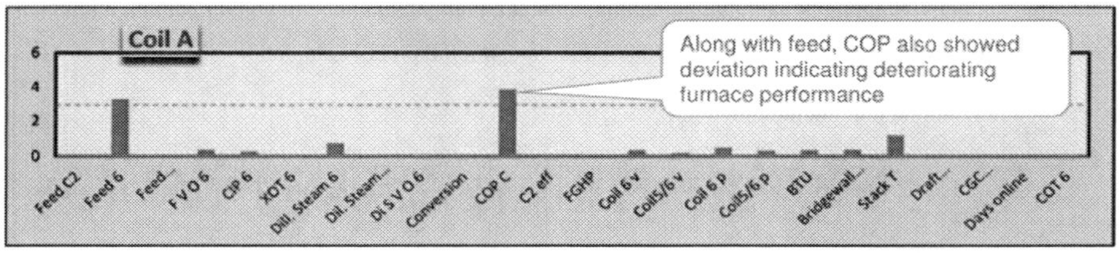

- ✓ PCA model predictions on ~25th day of furnace operation indicated:
  - ➢ Higher rate of TLE fouling, which may affect furnace run length
  - ➢ Deviation in coil feed flow possibly affecting coil performance/coking
- ✓ Based on further analysis, it was identified that drift in indicated coil outlet temperature (COT) which controls the coil feed flow was causing higher coking
- ✓ Instrumentation error in COT indication was corrected, additional actions like heat flux balancing was carried out
- ✓ Facilitated in extending the furnace run length considerably
  - ➢ Differential coking could have restricted run length to 35 – 40 days
  - ➢ Early identification and corrective actions helped in extending furnace run length to 55 days

INGENERO

Furnace Predictions using Partial Least Squares (PLS)

Statistical models based PLS technique is used to predict key furnace parameters to assist identifying deviations precisely and aid furnace performance improvement

- Key features
  - ✓ Partial Least Squares (PLS) is a powerful multivariate predictive technique
  - ✓ PLS fits equations between dependent and independent variables
  - ✓ Coefficient indicates which variable impacts predicted variable the most
- Furnace predictions
  - ✓ Model predicts critical furnace parameters such as COT, Feed flow, XOT, COP, CIP etc. based on historical furnace data
- Benefits of PLS models
  - ✓ Finding the deviations from normal/best operation accurately
  - ✓ Early detection of faults, equipment reliability
  - ✓ Assist in root cause analysis and optimizing furnace parameters for given condition

INGENERO

Furnace Reliability Matrix Example – An Excerpt

Sample reliability dashboard generated from analysis using the Technology tools, enabling identification and focus for actions

Furnace #5

15/Ethane	COT	XOT	TMT	XOP	TLE DP	TLE Outlet Temp
Today's	1558.6	1308.5/Coil1	1984.9/Coil4	46.7/Coil3	7/TLEB	730.5/TLEB
Deviation Yesterday	0.5	-0.9/Coil2	20.2/Coil3	0.9/Coil3	0.4/TLEB	0.8/TLEB
Deviation 10 Days	-2.49	-2.1/Coil5	63.6/Coil4	6/Coil3	2.5/TLEB	2.3/TLEA
Deviation SOR	10.2	-9.8/Coil1	142.6/Coil4	15/Coil3	4.5/TLEB	20.6/TLEB

Conversion (%)	64/64.2	SCR (°F)	789(E)/22.3/776.5(E)	Feed CV (%)	49.3/Coil 4
Dilution Steam CV (%)	81.2/Coil 5	No of Moles	2283.7/369.8	1500 # Steam Temp (°F)	910.0
BTU Firing (MMBtu/hr)	195.7/-0.18	NOx at Stack (ppm)	5.35/12	Coil Feed Imbalance (%)	5/5

Normal Operation ✓ | Deviation from Normal | Needs Attention | Abnormal Condition | Immediate Action Required

Matrix Deviations

Computed based on:

1. Difference between predicted and actual values
2. Difference between actual value and pre-defined limits based on furnace run length

Matrix Predicted Values

Predicted value is calculated for key furnace parameters viz. COT, XOT, CIP, TMT based on models calibrated by historical plant measurements and maintained to reflect current conditions.

Matrix Variable Limits

While furnace operating parameter limits are defined and set to be automatically cross-checked with, as part of the matrix computations and flagging/alert, some parameter limits within the matrix have been set variable to the furnace run length. This is because the significance of certain critical operating parameters like CIP, TMT, TLE ΔP etc. changes with days of operation.

INGENERO

LP Models : Optimizer

Based on an objective function (profit/production); customized LP models provides optimal operating parameters settings and distribution of feeds among multiple furnaces

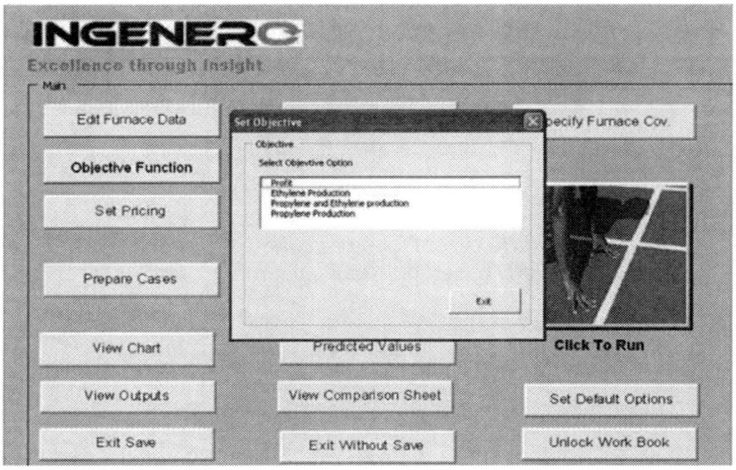

INGENERO

Furnace Dashboards

Analytics coupled with bird's eye view dashboards, provide a daily view with context for identification of problems and enabling of action

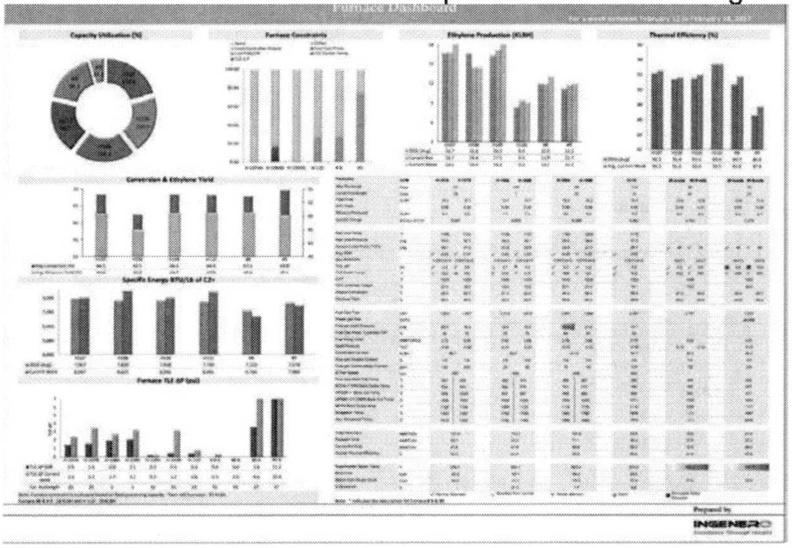

INGENERO

Conclusion

Application of big data analytics combined with IIOT through multi-model techniques and systematic analysis of furnace operation improves furnace performance, reliability and run length

- Furnace analytics greatly helps tap the best available capacity provided by the furnace design
  - ✓ Maximizing production
  - ✓ Maximizing run lengths or optimizing between run lengths and how hard to run the furnace
  - ✓ Optimizing yield
- There is no single tool available that can be used to model and understand the furnace operating idiosyncrasies – hence use of multiple tools to model the furnace, used in conjunction is require
- While effective multi-modeling is key, it has to be combined with the process of data collection, transforming it into information and knowledge using the models, visualization and dashboards to help interpretation and early detection and then an implementation and tracking process to ensure benefit realization

INGENERO

Questions?

INGENERO

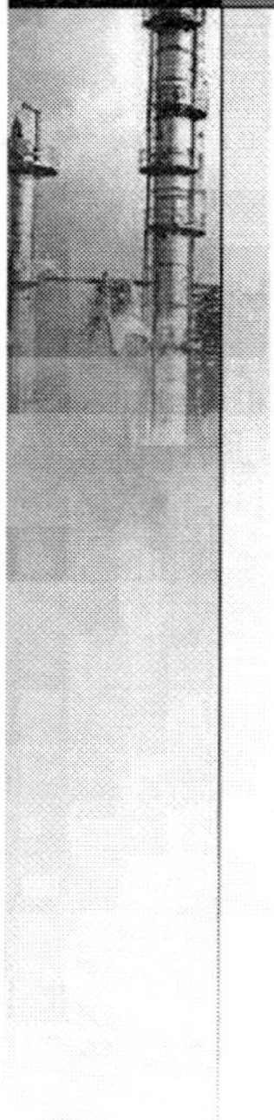

INGENERO

Excellence Through Insight

4615 Southwest Freeway, Suite 320,
Houston, Texas, 77027, USA

www.ingenero.com   info@ingenero.com

INGENERO

Confidential & Proprietary   23

485750 Smart Manufacturing Panel Discussion

Abstract:

Smart Manufacturing enables all information about the production/manufacturing process to be available when it is needed, where it is needed, and in the form it is needed across the entire manufacturing value-chain to power smart decisions. Learn about future of production and manufacturing from Smart Manufacturing experts. Hear firsthand experiences about setting strategies, supporting technologies, and lessons from implementation.

480429 A Data-Driven Early Warning System for Mining Accidents

Abstract:

Often, safety condition of a mine deteriorates before an accident happens. For example, before the 2010 Upper Big Branch Mine disaster, one of the largest in the U.S. history, the mine displayed an alarming rising trend of safety violations. Similar building-up behaviors apply to other major mining accidents in the U.S. According to the Mine Safety and Health Administration (MSHA), on average, there are more than a hundred thousand citations and orders issued yearly. The MSHA accident and inspection databases are an untapped resource for safety analysis. Our model analyzes semi-structured data from MSHA, attempting to uncover a company's safety culture from its public regulatory records. We discuss preliminary results from logistic regression. This work has implications for developing a data-driven early warning system. In the future, we plan to extend this research to inspection data from the Occupational Safety and Health Administration (OSHA) and the Environmental Protection Agency (EPA).

479231 New Ways to Visualize and Consolidate Alarm Management KPIs

Abstract:

Alarm management. We have been dealing with it for the last two decades, but its concepts, despite of been being essentially simple, are still of difficult interpretation by the operators, plant engineers and especially by the managerial levels. One reason for this is that the main clients of alarm management information (operators, plant engineers and site managers) are, as everyone else nowadays, overwhelmed with too much information and subject to pressures to achieve goals and make quick decisions. Thus, it is imperative to translate the main KPIs usually calculated for alarm management systems in a way that is easy, for someone unfamiliar to automation systems, to understand. These KPIs are, according to EEMUA 191: KPI1 - average number of alarm per time, per operator, KPI2 - maximum number of alarm per time, per operator and KPI3 - percent of time that plant are in a burst situation. Our company has adopted a commercial application to deal with the huge amount of information generated by its alarm systems (about millions of messages per month) and calculate those KPIs. This solution has interesting features like a MS Excel® add-in and a web interface to generate graphics and summarize alarm information for reports. However, the daily acquaintanceship with the matter encouraged us to develop more efficient ways to display the alarm results. The objective of this paper is to present the evolution of our company when reporting and analyzing the alarm management situation of each plant. We are going to present how we evolved from just reporting the KPIs generated automatically by the Alarm Management Software we use to presenting a graphical reports that are of easy and fast interpretation by the plant staff and managers. We have developed a highly automated MS Excel® report that makes use of SQL queries and custom store procedures to generate the information required for analysis quickly, thus increasing the productivity of the engineers in charge of generating the monthly reports. Also, we are going to present how to consolidate alarm management KPIs (defined by operator) in plant KPIs. This came up as a requirement to include alarm management practices in the Company´s Quality Management and Production System. At the same time that it expands the alarm management visibility and importance to the company and all level positions it creates the challenge to meaningfully consolidate the data not only for individual operator consoles, but for different levels and over different time periods. We are going to present the questions we were faced with and the tortuous way we have taken (including the mistakes) to arrive to a reasonably answer.

479232 A Simple Way to Calculate Control Loop Service Factor Taking in Account Plant Conditionals

Abstract:

Process safety considers several layers of protection to avoid that our process achieves an undesirable and dangerous situation. One of the most important of these protection layers is the basic process control system (BPCS). Nowadays, this protection layers is usually performed by distributed control systems (DCS) or programmable logic controllers (PLC), where each layer is configured as a control loop. Nevertheless, having these control loops configured is not enough. It is also necessary that the loop is closed and performing its job well. Although there are in the market several systems that provide means of analyzing control loops, they compete with the advanced process control (APC) and Process Information Management Systems (PIMS) in the acquisition of data from the BPCS, requiring a dedicated OPC server to get data in the required frequency. Secondly, many of the indexes calculated by them are of difficult interpretation by people outside the automation and process control environment. Thirdly, the information needed to calculate many of that index are already stored in PIMS systems. We have learned by experience that if a control loop is helping the operators to do their job, they will keep them closed on its correct operating mode (AUTOMATIC, CASCATE, REMOTE CASCATE, etc.). Based on this assumption, we decided that one of the simplest ways to track the performance of our control loops was to calculate the percent of time that control loops work on their target operating mode. This KPI is called by different name by different companies. We have chosen to call it service factor. All the information needed to calculate it is already stored in our process management information system (PIMS). To calculate this KPI is simple but there is a complication factor for continuous plants: conditionals. As soon as we started presenting this KPI for the plant engineers, we started receiving tons of justifications of why some loops were having a poor service factor. The problem was that the plant operation is not a steady state process. There are many situations when operators must break a loop (e.g.: during a reactor regeneration, unit start-up, etc...). In this paper we are going to present a simple approach using MS Excel and some PIMS store procedures to automatically take in account these conditionals and how the control loops are managed.

2017 AIChE®

SPRING MEETING & 13TH GLOBAL CONGRESS ON PROCESS SAFETY

March 26–30, 2017
Henry B. Gonzalez
Convention Center
San Antonio, TX

17 AIChE Spring Meeting, San Antonio, TX — 13TH GLOBAL CONGRESS ON PROCESS SAFETY

---

17 AIChE Spring Meeting, San Antonio, TX — 13TH GLOBAL CONGRESS ON PROCESS SAFETY

Empowering Human Capital for Improved Asset Reliability

Tim Olsen
Emerson Automation Solutions

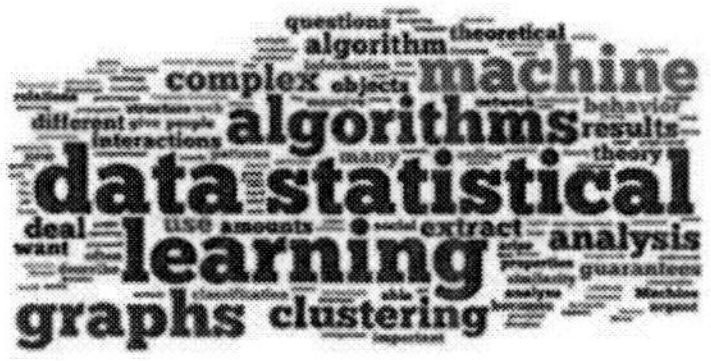

EMERSON.

Tim Olsen

- 27 Years in Refining Industry
  - 19 years Emerson Process Management as Technical and Business Consultant for Refining Industry
  - 8 years UOP as Technical Advisor on Startups
- Active with AIChE
  - AIChE Fellow Member
  - Fuels and Petrochemicals Division Past Chair
  - 2014 Spring Meeting Overall Program Chair
  - Current committee member:
    - Chemical Engineering Technology Operating Council
    - Center for Energy Initiatives Advisory Board
- Active with AFPM (*prior name NPRA*)
  - Operational Planning, Control & Automation Technologies Committee
- Active with Emerson Global Users Exchange
  - Board of Directors

Tim.Olsen@Emerson.com

Agenda

- Refinery Staffing
- Consumer Use of Smart Technology
- Why do these new technologies matter for the refining industry?
- Technology Overview
- Empowering Human Capital
- Value - Proofs

Fewer Staff in Refineries

- Technology has improved efficiency, thus less staffing requirements
- Aging staff retiring, and not always backfilled
- Fewer people want to work for the refining industry – negative PR
  - Can be dirty
  - Some work outdoors in adverse climate (hot, humid, raining, cold, winter, etc)
  - Can be dangerous (hydrocarbons can burn or explode, other chemicals such as acids and bases, etc)

Modern Technology: more control loops per console operator

US Oil Refining Industry Employee Productivity

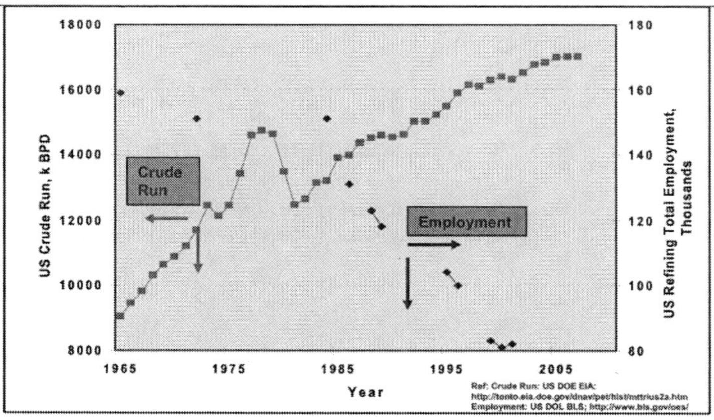

Employee Productivity Up 300%, Even With:
- Increased Complexity - More Crudes/ Product Grades
- Increased Regulatory Requirements

Sounds Good? Less People with Higher Productivity…

- Then why are some refinery staff working 10+ hour days?
- Why are there still unscheduled slowdowns and shutdowns?
- Why still a significant amount of manual data collection and analysis?
- How much data collected is still on paper or isolated electronic databases that make it difficult to access?

Unfortunately, despite advances in technology, some refineries have not changed their behavior… performing manual work that requires additional staffing that is no longer available.

Consumer Use of Smart Technology

Processing facilities like refineries follow advances in consumer technology

Data Example: Edyn- Smart Gardens

- Sensor - Monitors soil moisture and nutrition (conductivity), sunlight hours, humidity
- Connects to app on smart phone that accesses local weather forecasts to determine how much water is needed for garden
- Smart phone app connects via wifi to smart valve that controls water to garden based on current and forecast conditions – if rain is forecast it defers watering
- Can manually control watering remotely via your smart phone
- Provides recommendation on soil nutrition based on specific plants in garden
- Solar rechargeable batteries in sensor and valve

Source: https://edyn.com/

What if You Didn't Have to Take Your Car In For Service?

Car

- Must periodically be sent for service
- Critical components inspected
- You can't use it
- What happens between inspections?

BMW 7 Series (with ConnectDrive)

- Continuously monitored with car data and analysis sent to the dealer of your choice
- Early warning of problem with critical components
- Dealer contacts you if maintenance required on critical item
- Less disruptive
- Greater availability and performance

Technology Evolution Continues

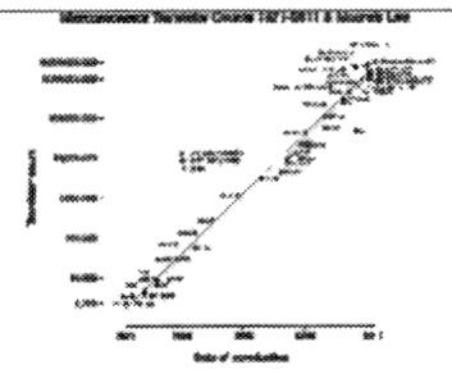

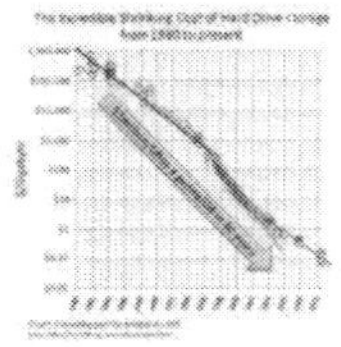

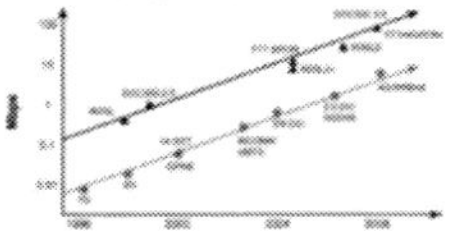

More Powerful And Smaller Computing

Cheaper Data Storage

Faster Communication

Data Is Increasing

- Walmart's Online Database
  - More than 250 billion rows
  - More than 20 million updates per hour
  - 3 years history available online
  - 24,000 queries per hour from 1500 users
- Internet Devices
  - Gardner group estimates 5 billion connected devices in 2015 – 25 billion by 2020
  - Plant device connectivity is increasing similarly
- Process Company
  - 10 million tags across 15 sites
  - Implementing system to provide real time access to 3 years of 1 minute sampled historical data for tags
- Refining Company
  - Reported total of 80 billion data items from four sites in one year
- Oil&Gas Company
  - Reported 1 Petabyte (10^{15} bytes) "Data Lake" of company data created to facilitate 3rd party cooperation/ coordination

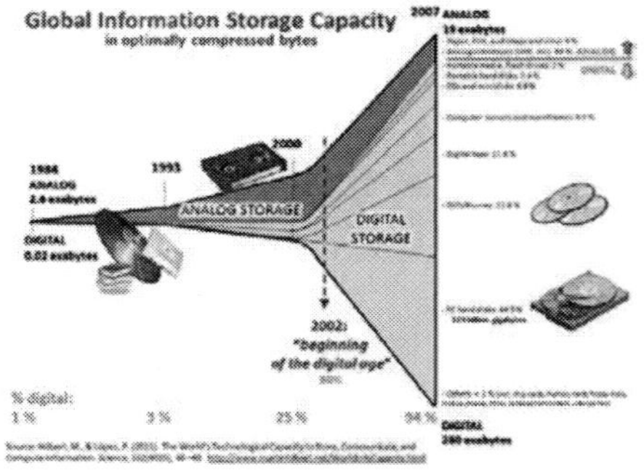

Data Is Increasing

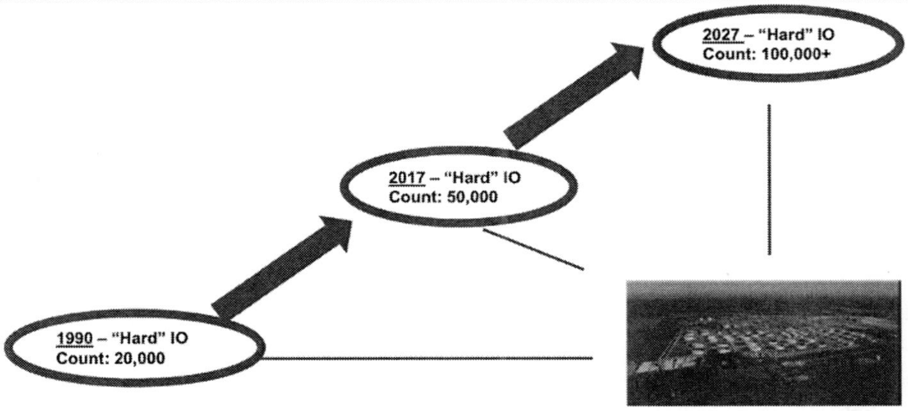

"Single-Train" Refinery

Photo Source: US DOE

Is More Data Better!

Why do these new technologies matter for the refining industry?

What Do These Technologies Mean For The Refining Industry?

You can measure and collect data from almost anything/ everything

You can store all data with fast/ cheap access

You can send the data anywhere

Combining data with analytics can significantly improve operations

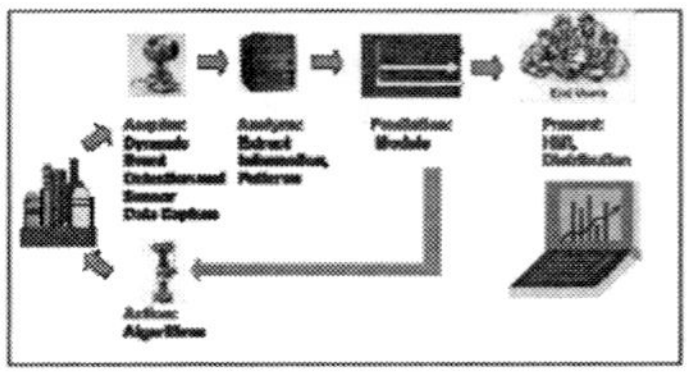

Creating Value From Data - Looking Forward

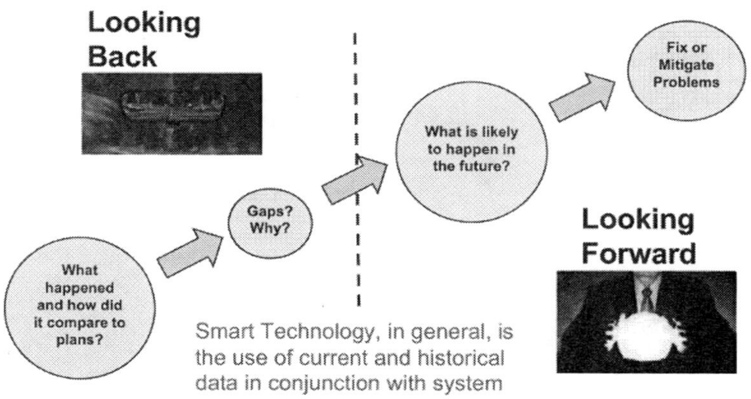

Smart Technology, in general, is the use of current and historical data in conjunction with system models to predict future trends in real-time (near).

Technology Overview

Smart Technologies Architecture

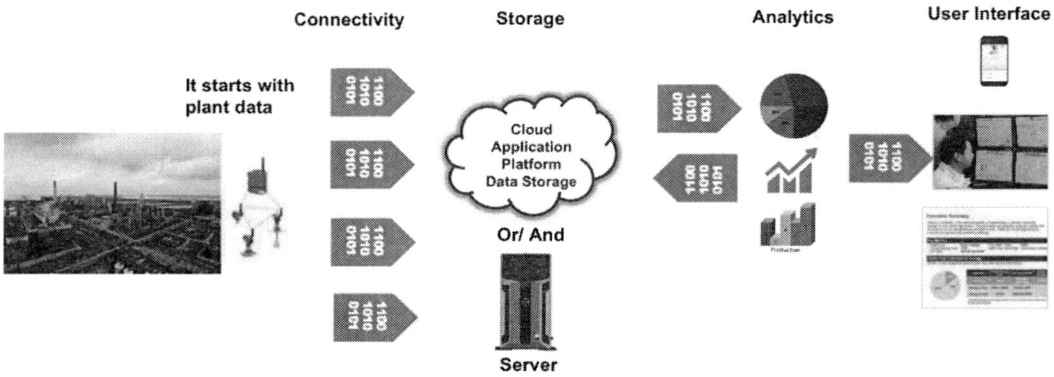

Types of Analytics

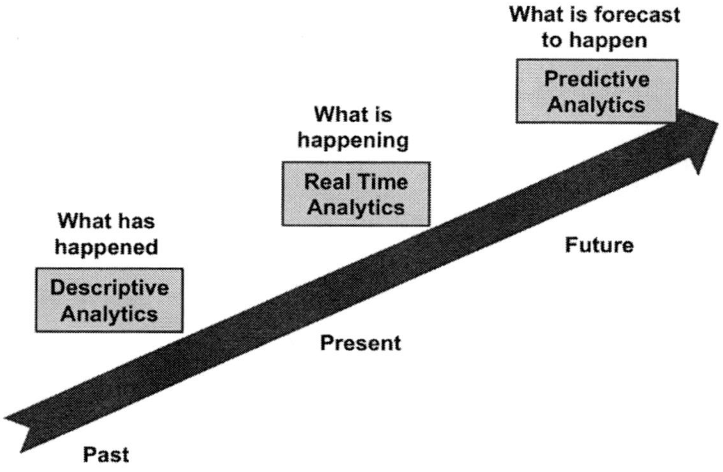

Analytics Software Distill Raw Data to Information and Knowledge

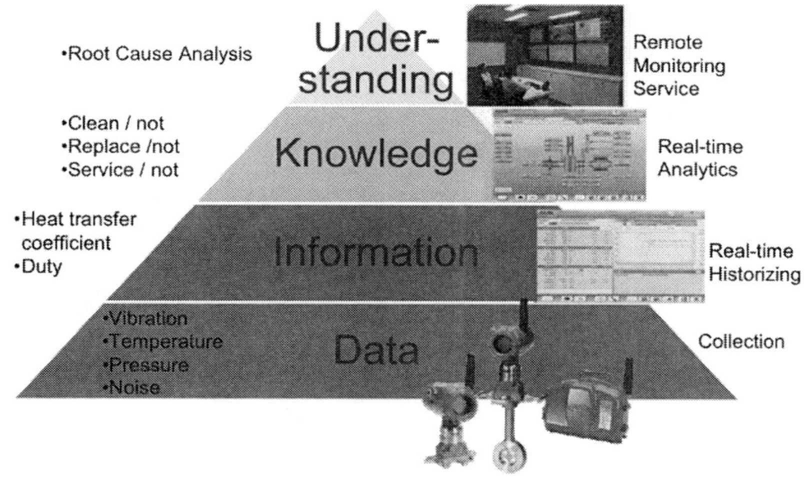

The Industrial Internet of Things

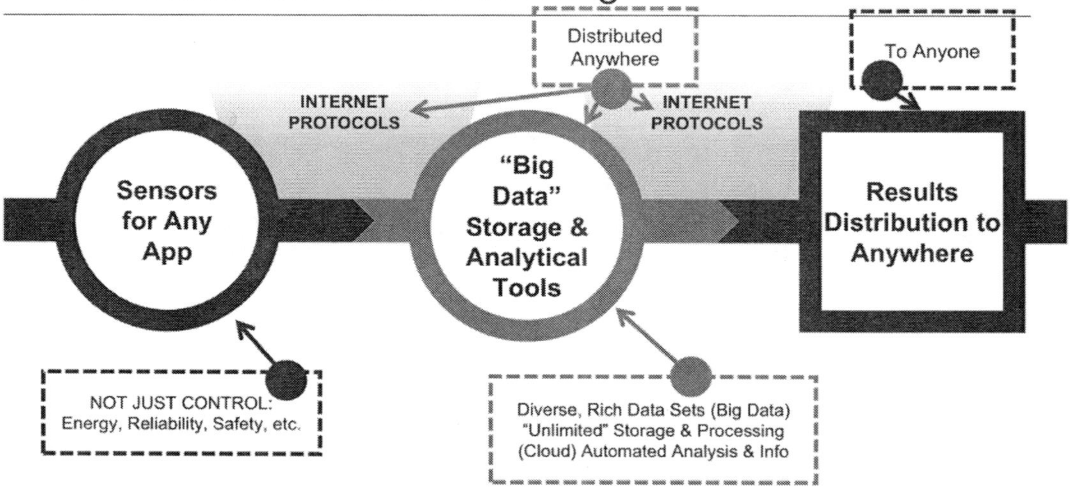

Internet Enables Remote Monitoring of Equipment

- Enterprise level
  - Global technology center of excellence
- By equipment manufacturer
  - Pump, valve, ACHX, CT, etc.
- By third-party plant services provider

IIoT is a group of enabling technologies

IIOT Connectivity – WirelessHART - Proven Technology

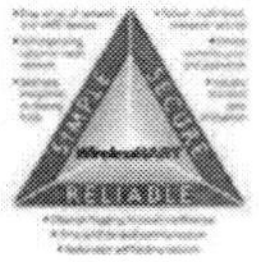

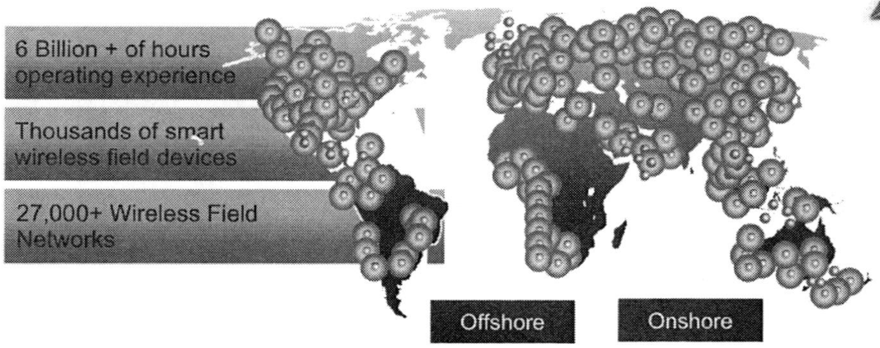

Connected Plant Vision

Center of Excellence
- Your own
- Or Connected service

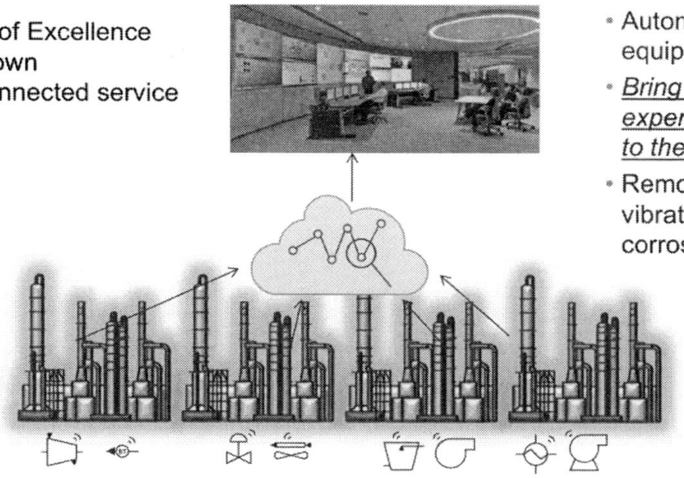

- Automatic collection of equipment data
- *Bring the data to the experts; not the experts to the data*
- Remote pool of experts: vibration, valves, corrosion, steam traps...

Typical Refinery Smart Technology Implementation

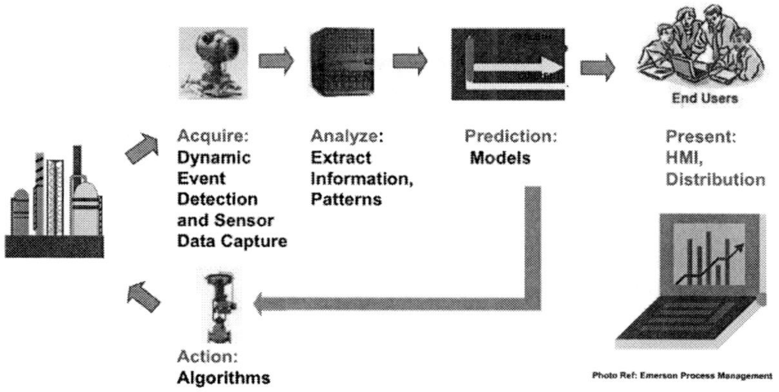

Move from reaction after disruption to planned maintenance prior to disruption to anticipation and optimization

Empowering Human Capital

Empowering Human Capital

- Technology is just the beginning
- Adding Sensors will produce more data, and more challenges if the data is not automatically analyzed
- Analyzed data provides information
- Timely information can provide alerts to appropriate staff

- However, staff should be trained to act in a timely manner to correct abnormal operation or prevent imminent failure of an asset
- Many times SOPs will need to be modified to ensure proper behavior based on this new information

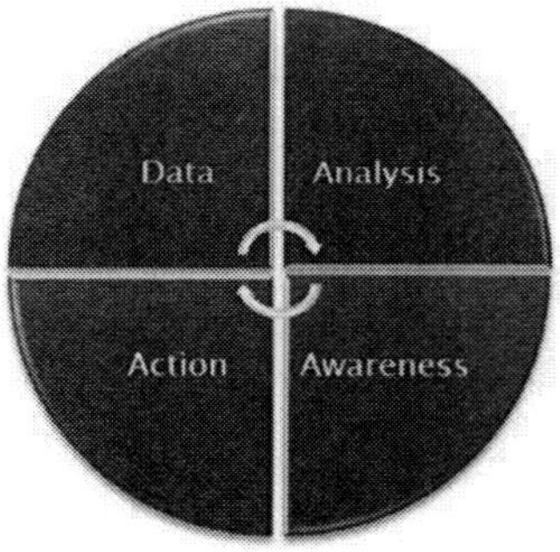

Change in Behavior – Value Realized

- Traditional
  - Collect and historize data
  - Look back and evaluate data *after* incident
- New Approach – Predictive Analytics
  - Automatically analyze data and turn into information
  - Look forward and alert *before* abnormal operation or imminent failure
  - Perform appropriate timely action to avoid asset failure

Creating Value From Data

Smart Technology Impact on Operations

Use the same basic approach we control processes with- Feedback Control

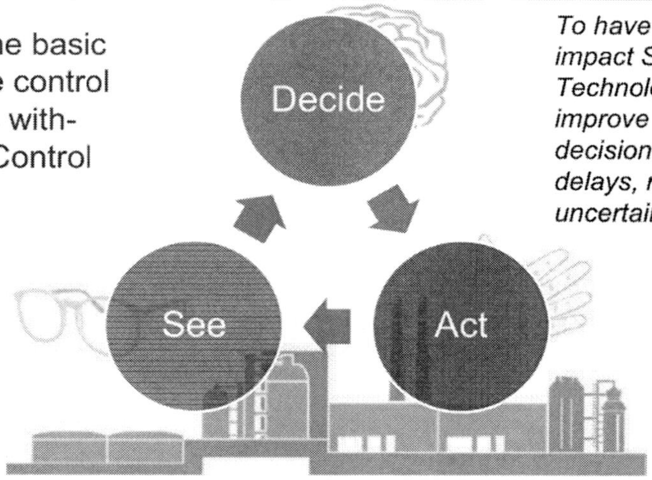

To have a financial impact Smart Technology has to improve the plant decision cycle – reduce delays, reduce uncertainty, etc.!

3

Technology by Itself is Not Enough

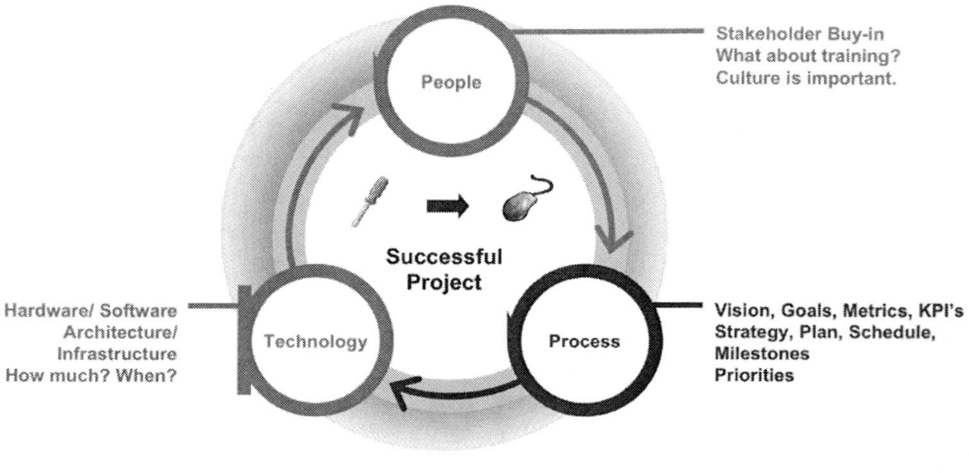

3

Case Study: Smart Technology & Maintenance Improvement

- Significance
  - Saudi Arabian Petrochemical site that is considered a world-class complex with 4 large integrated plants
  - Successful implementation of complete integrated maintenance program
  - Built on installed base to create multi-phase asset management program
- Project Scope
  - Implementation of Asset Management including Reliability Centered Maintenance on 6,000 assets
  - Uses predictive analytics and diagnostics from field instrumentation
- Audited Value
  - 12% decrease in maintenance costs
  - 2% increase in Plant availability

http://www.emerson.com/resource/blob/38018/e0ae5996c23b97c284269336df16ba06/petrochemical-complex-ratchets-up-reliability-data.pdf

Case Study – Delivering Strategic Business Value

- European Refiner – Five Sites
  - Implemented Smart Technology over past three years
- Results
  - Increased white product yield by 2.9% (basis crude)
  - Reduced hydrocarbon losses by 15%
  - Increased operational availability by 0.6%
  - Reduced energy usage per ton by 2%
- Estimated margin improvement of $167 million per year (across all sites)
- Target for next two years is an additional $ 250 million per year of margin improvement

3

Sustaining The Program

Keys To Long Term Program Success

- Accountability – Need To Have A Specific Individual Or Group Responsible
  - Need Top Management Sponsor And Support
- Visibility – Need Regular Reporting Of Results And Trends To Plant Management ($/ Yr Savings)
- Monitoring – Need Automated Calculation Of Equipment, Unit, And Site KPI's
- Training staff to effectively utilize the new information… an alert of abnormal activity should not generate confusion

3

Summary

- Smart Technologies are continuing to evolve with many potential applications in the process industries
- Implementation of Smart Technology in the process industries has distinctive requirements/ issues due to the data character and data
- Current applications in process fault detection, availability, safety and optimization have proven value – there will be more in the future

- As technology and analysis advance, the human element cannot be forgotten.
  - Staff needs to be trained on how to effectively use the new information
  - SOPs may need to be modified to change behavior

3

17 AIChE Spring Meeting, San Antonio, TX

13TH GLOBAL CONGRESS ON PROCESS SAFETY

Empowering Human Capital for Improved Asset Reliability

Tim.Olsen@Emerson.com

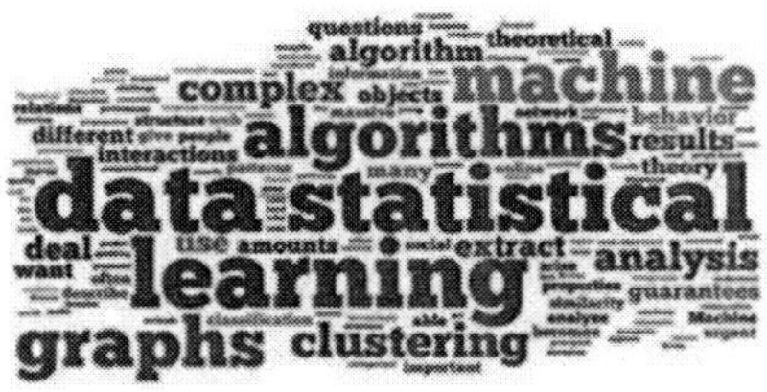

EMERSON.

2017 AIChE® SPRING MEETING & 13TH GLOBAL CONGRESS ON PROCESS SAFETY

480299 Smart Plant Applications in Crude Distillation Units

Abstract:

Smart Plant represents an approach to delivering optimal performance of plant assets. It focusses on the intelligent use of plant data and predictions of performance to make decisions that deliver operating benefits. These can be classified in terms of economic, environmental, safety and quality. Crude units form the heart of an oil refinery and their operation drives overall refinery profitability. Dwindling margins and tighter regulations make the adoption of new technologies and concepts even more important, so operators can maintain any possible competitive edge. This paper presents our ideas and experiences of Smart Plant applications in refinery crude distillation units.

Smart plant is a form of intelligent manufacturing. There are a number of key elements within a Smart Plant system that need to be considered to ensure smart decisions are made.

The Smart Plant approach requires a sensor network. The associated instrumentation provides the data landscape from which conclusions are drawn regarding performance. However, such data sources are prone to uncertainty, for example as a result of broken sensors, faulty instruments, signal drift and breaks in the data chain. Data quality is measured in terms of frequency of records and uncertainty.

Data values need to be validated and reconciled. There are several methods to doing this, and our approach has centred on using State Estimation technology via a Kalman filtering approach. Based on a high-fidelity, mathematical model of the process, this approach enables appropriately accurate estimates to be made of unknown variables by estimating probability distributions for measured variables. PSE's gPROMS ProcessBuilder environment based on the equation-oriented gPROMS platform is used to facilitate such an assessment. Dynamically changing data are gathered and a full set of operating parameters can be calculated which describe the current process operation.

In fact, this paper shows how State Estimation is applied in real time. The key principle is to have the detailed physical model of the process to provide an integrated environment and then conduct process simulation trials.

With the data now validated and a time-averaged set of performance operating parameters defined, the model can proceed and the process optimised, generating set-points that can be fed to the control system directly or to an APC application. Hence, model-based Supervisory control is achieved.

We also now highlight a further objective of Smart Plant applications - multi-focussed optimisation. The scope of the optimisation can be steady-state focussed, minimising overall cost, emissions or other parameters such as flaring. We can also perform dynamic process optimisation - for example how to most efficiently transition from one operating state to another, for example as feedstock changes.

Another objective of the Smart Plant approach is asset optimisation, where design, operations and maintenance aspects are all considered in the analysis. This topic is achieving more focus now because of the realisation that, an optimised system will very quickly become inefficient if it goes off-line, for whatever reason, and production is reduced or stopped. Hence, the attributes of maintenance programmes must be considered.

Finally, the way the results of Smart Plant applications are disseminated is very important. Pushing information to stakeholders and visualisation of performance through Key Performance Indicator (KPI) tracking is very helpful to adopting such improvement advice.

An industrial case study is presented of our experience using the gPROMS platform to optimise crude distillation units. It shows how operating costs are optimised alongside maintenance costs, in real time, to provide an overall optimised crude unit. The example uses real-time data acquisition and state-estimation techniques to ensure data integrity and informed decisions. Operators then make decisions on process improvements with confidence, realising significant cost savings in the process. The results are presented to users via data visualisation tools and KPI dashboards, ensuring appropriate dissemination of results and realisation of the benefits.

481919 Regression Strategies for Large Data Sets (poster version)

Abstract:

Plant operators and engineers are increasingly using historian data to gain insight into the relationships between and amongst process parameters and product attributes. A universal goal is the determination of the operating conditions that produce the highest product output and/or quality per unit cost. Models provide insight, but seldom account for myriad physical nuances which sometimes exhibit significant influence.

The quantity of data that must be analyzed in pursuit of the sought inferences is invariably large, consisting of hundreds or thousands of quantities sampled at many millions of points in time. It is seldom possible to process data sets of this size in computing resource memory.

A number of frameworks for partitioning large data sets have been developed and popularized, though adoption by industrial companies has been notably slower than by IT-centric enterprises. This paper endeavors to demystify the processing of out-of-memory data for an engineering audience.

Regression, which is fundamental in data analysis, is used to motivate use of the techniques. This paper begins with computing simple statistics. Subsequently, some strategies for matrix manipulations (relevant to both direct and iterative methods) are discussed. Finally, an illustrative example of a large data set regression is presented.

Agenda

- Introduction
- Industry Challenges
- Teradata Approach
- Teradata Solutions
- Teradata differentiators

2

TERADATA

IoT – What's New?

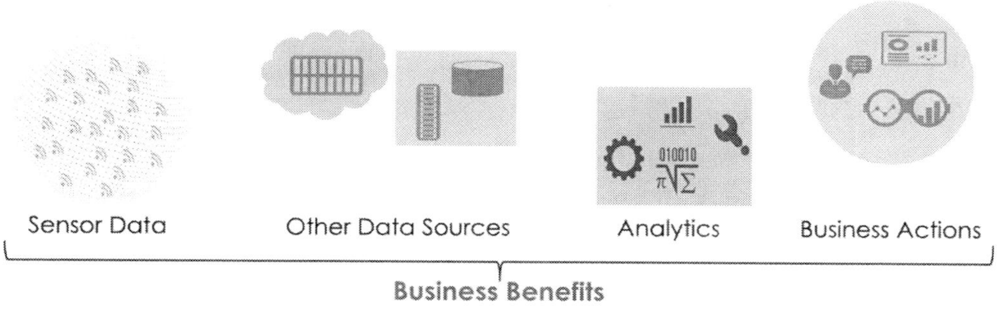

Right data, right insights, right actions driving business outcomes – revenue growth, customer satisfaction, operational efficiency

3 TERADATA

What are Industry Leaders Saying

- **35%** of US manufacturers are currently collecting and using data generated by smart sensors to enhance manufacturing/operating processes
- Organizations capture but face challenge using data. **48%** spend the most time reviewing data for quality and consistency

Business Focus for IoT initiatives

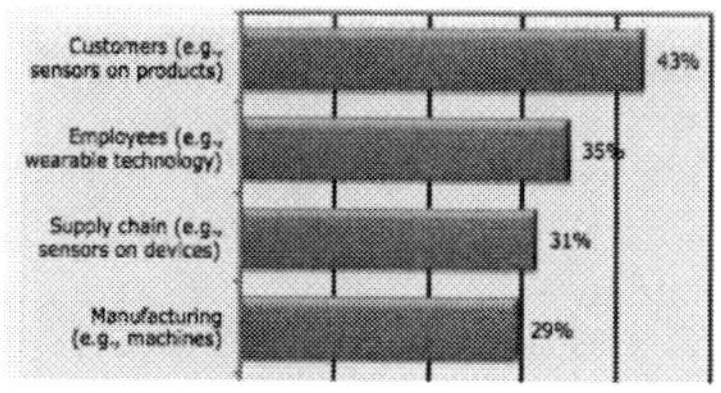

Issues with Implementing Technology

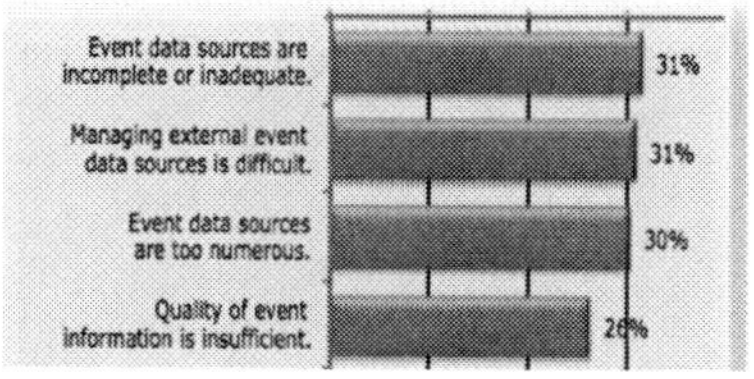

4 TERADATA

Today's Challenges – Data Issues

It comes from

Fleets: from lots and lots of similar things

Systems: across the same big "thing" or facility

Collectors: "big models" or monitoring

It can contain...

Outliers: Which of my things are behaving differently?

Emergent behaviour: Is my system changing to a new state?

Events: are there hidden signals?

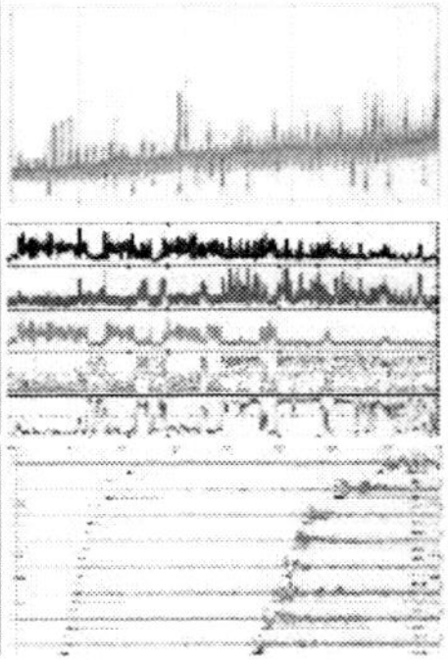

It has impact

"Fleet-wide" 24/7 for holistic management

High-level KPIs at business units and facilities level

Performed at sub-second level and data kept for decades

TERADATA

Today's Challenges – Analytics Complexity

Mean Shift
Neural Networks
Affinity Analysis
Spectral Clustering
Feature Selection
Random Forest
nPath Analysis
Association Analysis
Nearest Neighbors
Support Vector Machines
Time Series Analysis
Geo-Spatial Analysis
Machine Learning
Clustering
K-Means
Regression
Dimensionality Reduction
Decision Trees
Scoring Models
Graph Analytics
Text Analytics

- Analytical Tools
- Model Generation
- Validity
- Accuracy
- Speed
- Scalability
- Repeatability

TERADATA

Today's Challenges - Business link – Operationalizing of Analytics

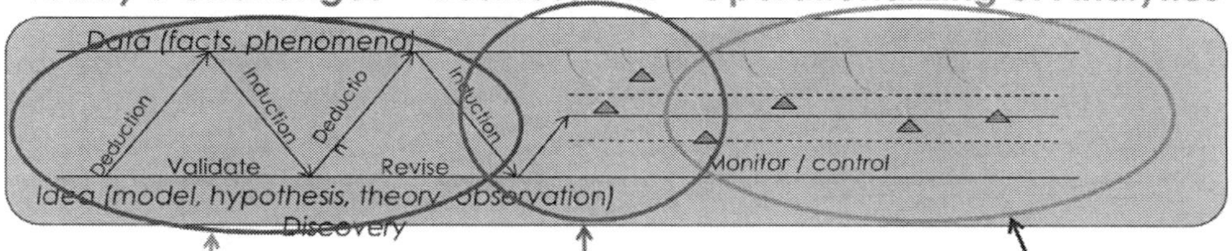

Engineering Focus

- Selecting right tool based on business need
- Setting up a scalable and repeatable environment
- Difficulty in converting business need into an engineering requirement

Implementation

- Missing: Taking discovered relations into production – quickly, efficiently, accurately
- Integration with BI tools, Engineering tools, Reports
- Platform decisions

- Business Contextualization
- Deriving meaningful business actions
- Adaptability of business
- Ability to monitor and measure the impact

7 TERADATA.

In Summary, Your Business problem looks like

The Problem

OPERATIONAL SYSTEMS **DECISION MAKERS**

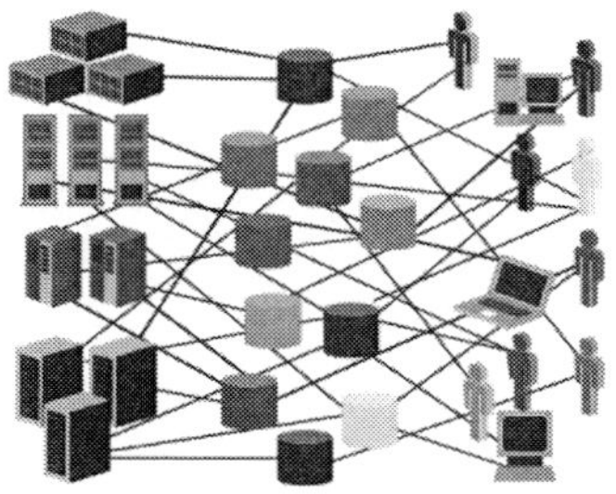

Proliferation of data marts resulted in fragmented data, higher costs, poor decisions

- Represent one silo of the business
- Bringing many silos of data can prove to be difficult
- Cannot keep pace with frequent changes in market conditions
- Decision by leadership is rarely made with complete data picture
- Creates poor communication in value chain
- Data Latency is technology or work process

8 TERADATA.

How can we provide decisions at the pace of business

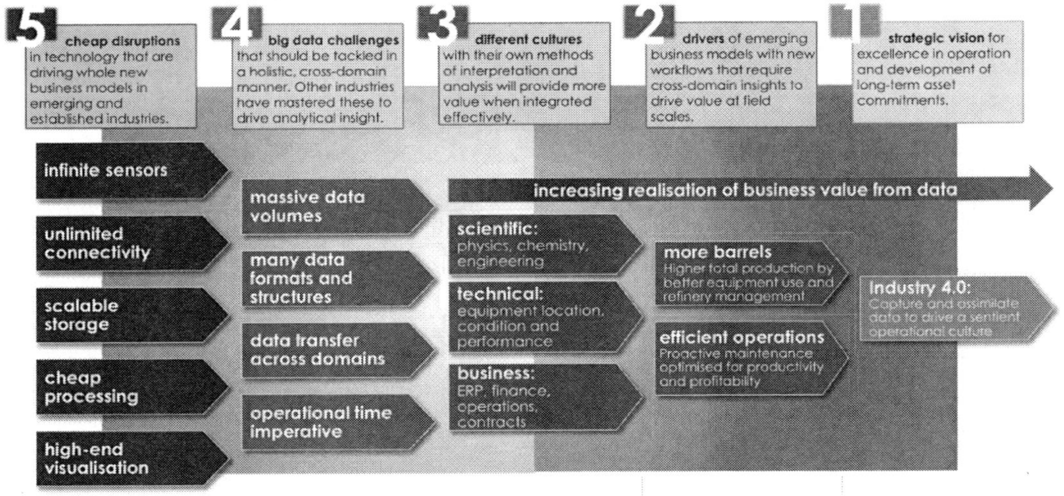

TERADATA

Teradata Approach & Solutions for Industrial IoT

AoT Approach – Overview

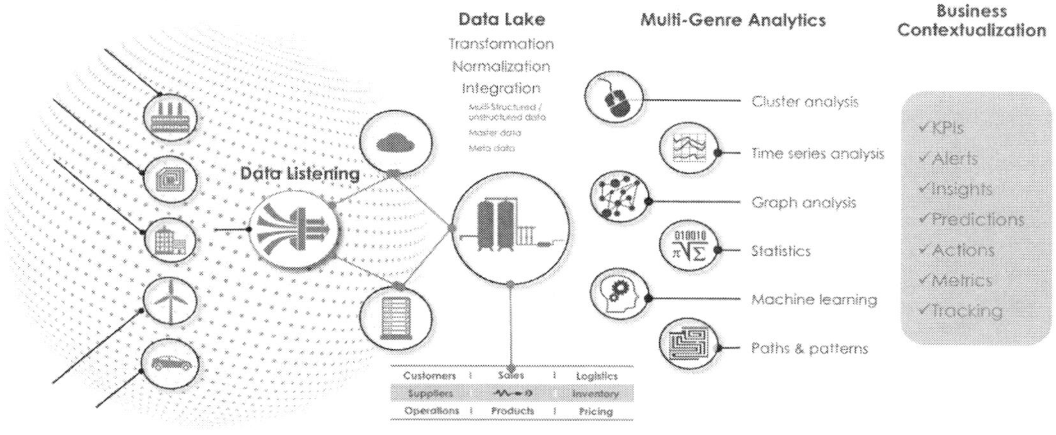

11

TERADATA

Oil & Gas Applications

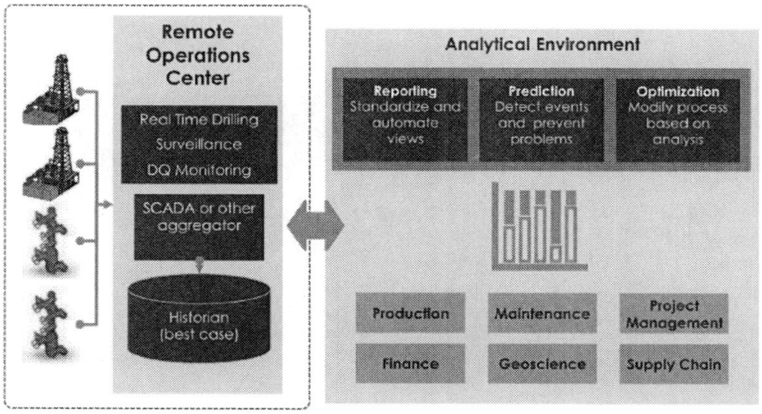

12

TERADATA

Contents

1. Sensor Data Qualification
2. Connected Factory
3. Smart Assets
4. Connected Supply Chain
5. Services 360

13

TERADATA

Solution: Sensor Data Qualification

Data Ingestion	Data Transformation & Storage	Data Analytics	Business Contextualization
DQ Stage 1	**DQ Stage 2**	**DQ Stage 3**	**DQ Stage 4**
Qualify Data Ingestion	Qualify Data Transformation	Qualify Data Analytics	Qualify Business Relevance

* DQ: Data Qualification

14

TERADATA

Contents

- Sensor Data Qualification
- **Connected Factory**
- Smart Assets
- Connected Supply Chain
- Services 360

15 TERADATA

Solution: Connected Factory

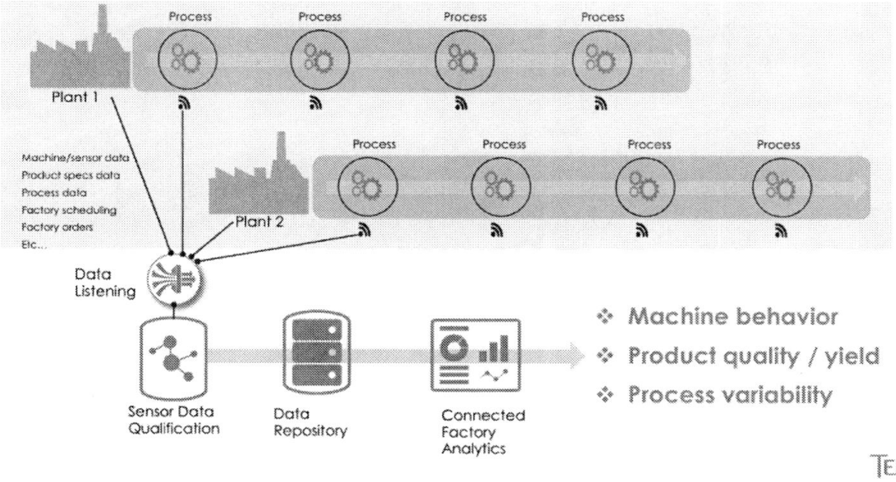

- Machine behavior
- Product quality / yield
- Process variability

16 TERADATA

Connected Factory – Machine Behavior

Monitor and analyze machine logs and machine sensors as time series to predict machine failures and prevent unplanned downtime

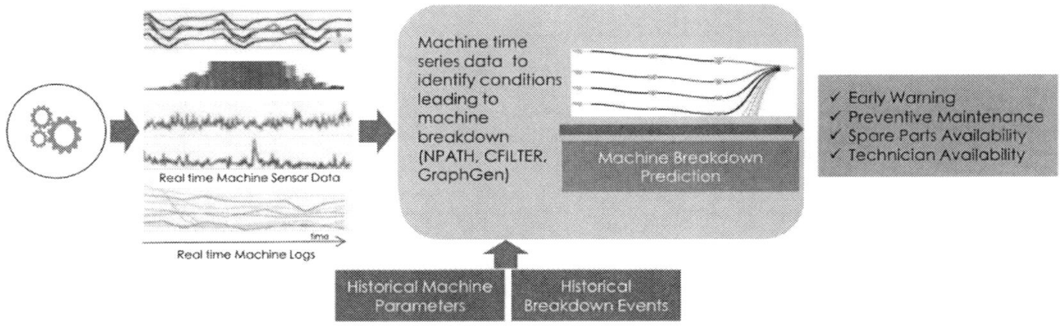

17

TERADATA

Connected Factory – Product Quality / Yield

Analyze product quality issues to identify conditions that lead to quality defects, yield issues, and build predictive models for detecting these issues

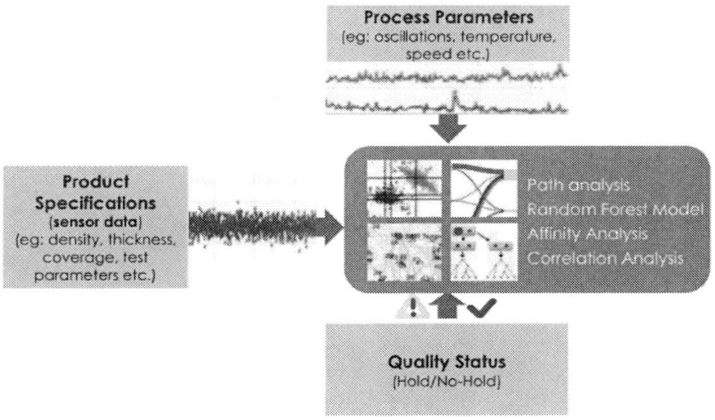

- ✓ Early detection of potential product quality issues
- ✓ Proactive adjustment of conditions that could lead to quality issues, or reduce yield
- ✓ Simulate yield / quality given a set of conditions (design of experiments)

18

TERADATA

Connected Factory – Process Variability

Analyze process variability impact to address process maturity issues, predict production yields/outputs, and reduce order commitment risk

Sub-Process Variability
(Process flow rules, Real time processing times)

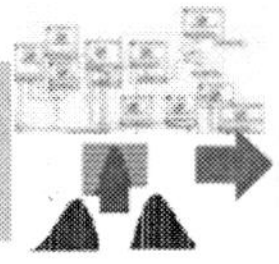

- Variability impact analysis
- What-if analysis & scenario modeling
- Alternate production path analysis
- Order risk analysis
- Raw material availability

✓ Increased productivity
✓ Faster time to process maturity
✓ Fulfillment predictions
✓ Risky order alerts
✓ Lowered work-in-process inventory
✓ Improved material availability

TERADATA.

Contents

TERADATA.

Solution: Smart Assets

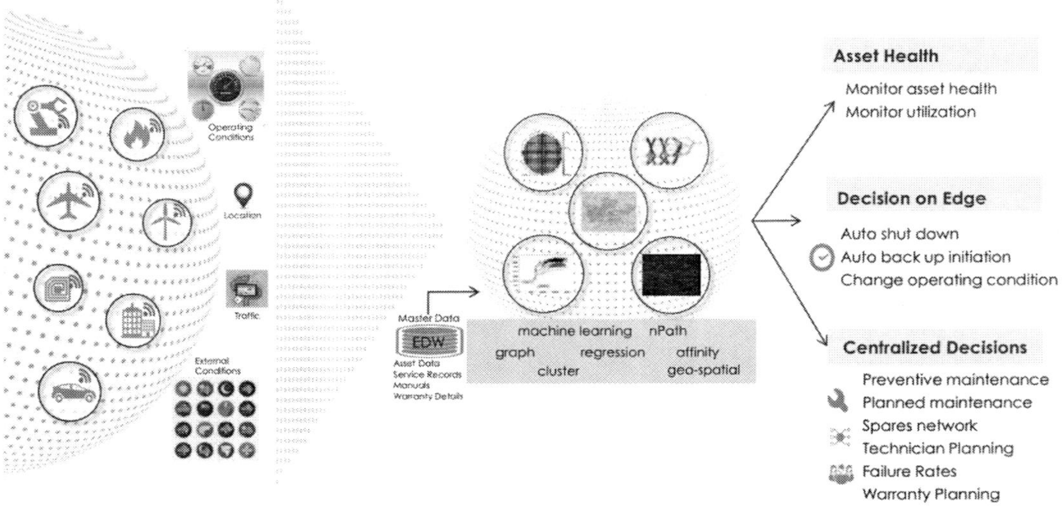

Smart Assets – Use Cases

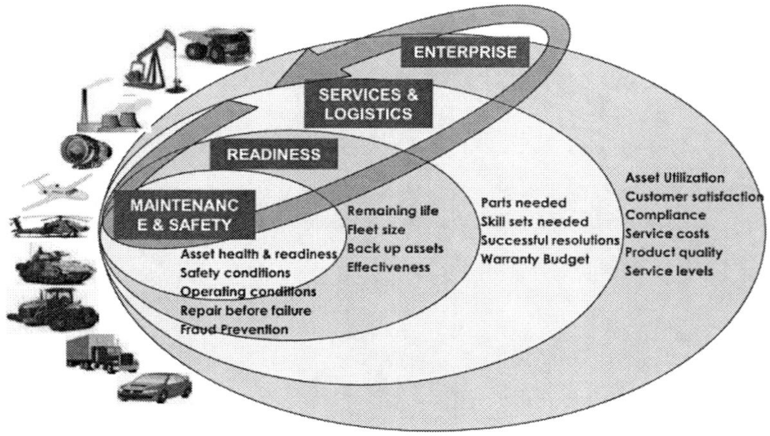

TERADATA.

Contents

23 TERADATA.

Solution: Connected Supply Chain

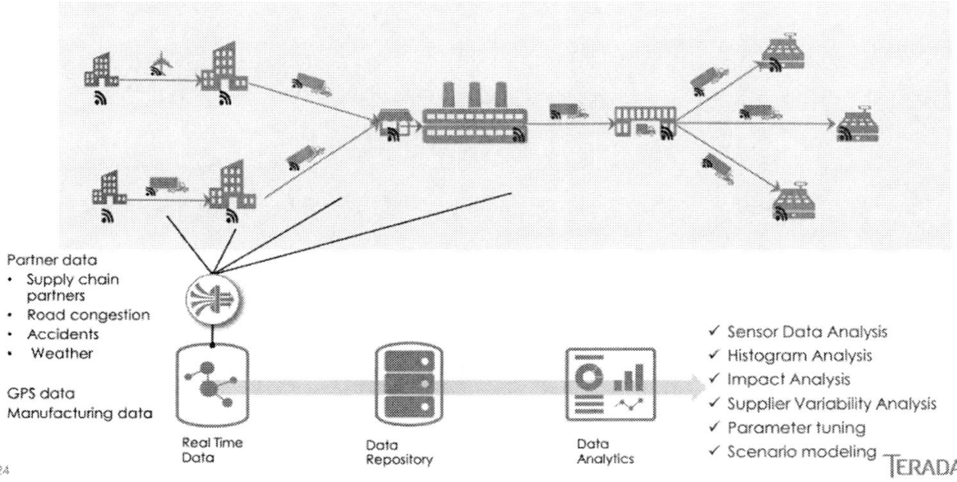

24 TERADATA.

Connected Supply Chain Maturity Map

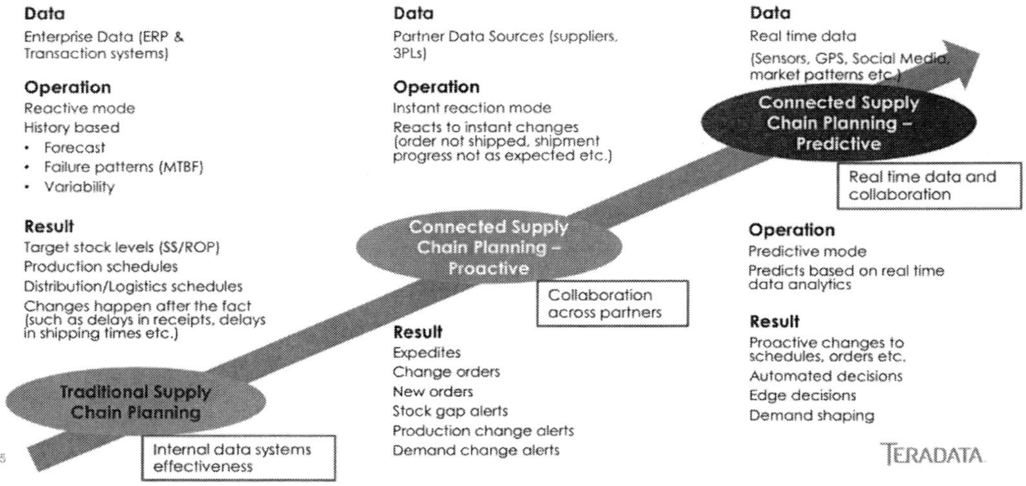

25 TERADATA

Contents

26 TERADATA

Solution: Service 360

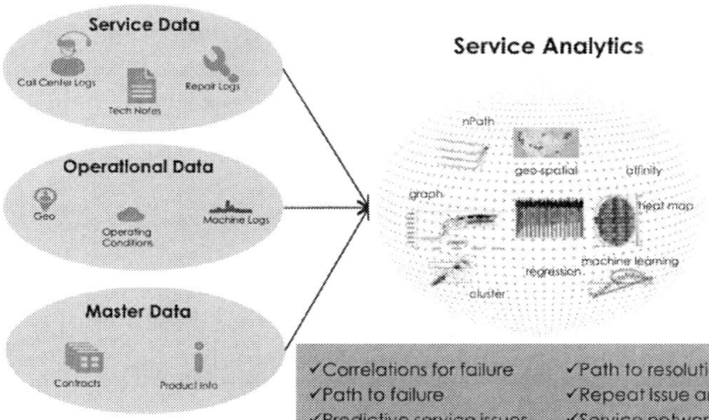

✓Correlations for failure	✓Path to resolutions
✓Path to failure	✓Repeat Issue analysis
✓Predictive service issues	✓Service network analysis
✓Early warning	✓Service profitability analysis

BI/Reporting
- ✓SLA Adherence
- ✓Comparative Analysis
- ✓Customer Patterns

Service Improvements
- ✓Resolution Guide
- ✓Product Quality
- ✓Proactive Fixes

Service Offers
- ✓Service offer redesign
- ✓Offer upsell
- ✓Offer cross-sell

Service Planning
- ✓Tech Resource Planning
- ✓Spares Planning
- ✓Service Network Planning
- ✓What-if Analysis
- ✓Capable to Service Analysis

Service 360: Business Benefits

- Proactive issue identification
- Faster & accurate issue resolutions

- Lower cost to serve
- Higher service profitability
- Increase in revenue through cross-sell and up-sell

- Higher net promoter score
- Customer satisfaction & retention

Sample use cases

- Organizations supporting data center temperature control and uptime
- Organizations leasing heavy equipment for production purposes (O&G, Mining)
- Organizations selling enterprise servers with tight support contracts

Enablement

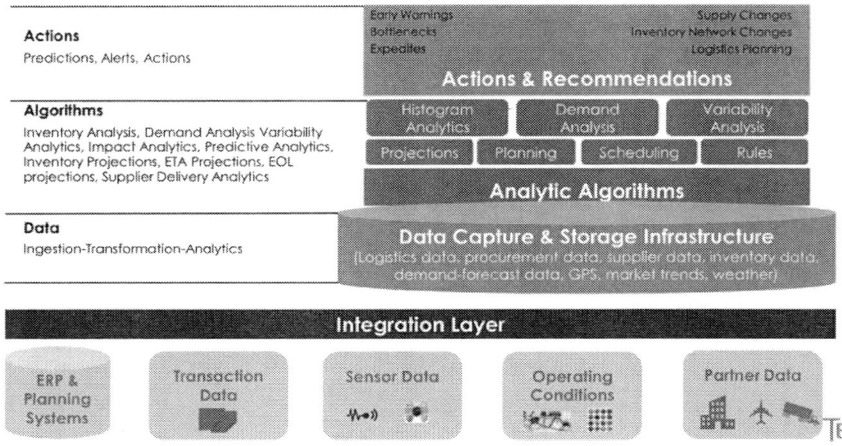

29 TERADATA

Migrate From Projects To Frameworks

Shift focus from a single, tactical exercise to a comprehensive framework

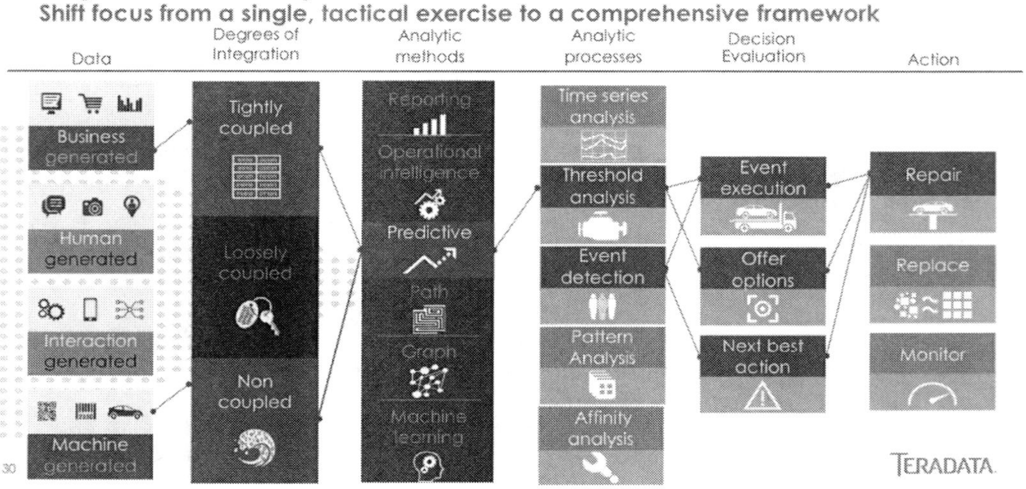

30 TERADATA

TD Differentiators

Outcome based

- ✧ Actionable recommendations
- ✧ Monitors key metrics
- ✧ Incorporates measure, monitor & feedback loops

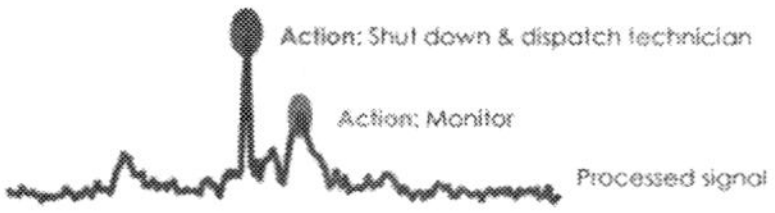

Standardized

- ✧ Specific business problems
- ✧ Proven Approach
- ✧ Data quality management
- ✧ Data models/dictionaries
- ✧ APIs/Analytical Models
- ✧ User Interface

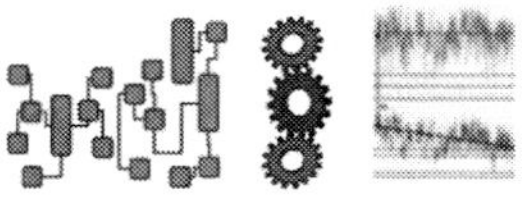

Configurable, Robust & Scalable

- ✧ Business rules can be configured
- ✧ Scalable environment
- ✧ Establishes a flexible environment to solve to other business problems
- ✧ Adapts to multiple technologies
- ✧ Highly Secure

TERADATA

Concluding Remarks

- You've learnt some new buzz words: Industry 4.0, IoT, Big Data, analytics...and hopefully you understand what they mean and how these ideas impact you
- You've learnt that there is a data driven approach that can build on and enhance your current processes
- You have a sneaking suspicion that it might be possible to unlock more value from your data
- Some of this unlocking could be straightforward integration of data sets but some of it will require some thinking to understand what the data might be telling you (Discovery)
- You have some good ideas on what business value and impact you can drive with this data

32 TERADATA

481782 The Canopy Platform for Process Monitoring and Optimization (poster version)

Abstract:

Enthought provides product and consulting solutions for the chemical, petrochemical, pharmaceutical, and semiconductor industries. At the core of these solutions is the Canopy Platform, which brings together a host of powerful components for data analysis, visualization, local and remote data access, and data exploration. Canopy Platform solutions allow engineers to easily browse and visualize huge amounts of data stored remotely, do interactive MATLAB-style analysis on it, and develop analysis or detection algorithms and deploy them at scale over large datasets using cluster or cloud compute resources. Detection and prediction algorithms developed in this fashion can be applied to monitor live incoming data and provide feedback to operators, used to provide input to maintenance scheduling, or used to debug production issues. These solutions place a strong emphasis on both convenient desktop user interface (useful for experts to develop their algorithms) and web dashboards for monitoring or data consumption by a broader audience. Web service components makes detection or prediction algorithms written on the platform accessible programmatically, so that the development effort can be reused by other systems and software throughout the business.

In this talk, we show how we leverage the Canopy Platform to build modern, customized software for analyzing plant process systems. We'll demonstrate the ad hoc analysis capabilities and workflow-oriented user interface that simultaneously simplifies open-ended and routine tasks. These capabilities make use of the extensive library of analysis routines that are available to the Canopy Platform and the data management component that enables uniform access to raw files, databases, and remote data stores. We'll walk through a cluster analysis workflow that can make use of a company's existing compute resources to scale to thousands of controllers and years worth of batch or continuous data. We will also describe the larger vision for this software as an automated plant advisor that monitors the status of processes, warns when parameters are out of expected bounds, forecasts failures, and offers solutions to those problems.

481781 Using Analytics for Process Safety (poster version)

Abstract:

Managing plants flooded with thousands non-statistically based alarms daily creates a major challenge for plants wanting to minimize risk. Operators are dismissing non-critical alarms throughout the day, but how many of non-critical alarms does it take to make a critical alarm? 35? 75? 1,000? (is this death by 1,000 cuts?)
Using real-time analytic monitoring can create a scientific way of filtering the data to provide early warning detection, while reducing the noise from fixed set point alarms which may not be relevant for the current process. This presentation will delve into a holistic approach for managing risk within the plant and throughout the whole organization. Monitoring feed stocks, in process parameters, and maintenance data using analytical tests creates an overall view of plant process stability. We will explore case studies and best practices for managing and guiding plant decisions creating a more safe environment for both assets and people.

481780 Making Cognitive Solutions Real Today -Stories and Examples (poster version)

Abstract:

IBM and others have aggressively invested in the advancement of Cognitive technologies such as Natural Language Processing, Machine Learning and Image Recognition, over the past decade . Cognitive driven solution are now having a profound impact on social and economic systems around the world. In this session, IBM will walk the audience through relevant use case and describe what's involved in designing, building and implementing cognitive solutions in a chemical engineering environment. The session will conclude a live demonstration of cognitive solutions.

481779 Practical Challenges and Opportunities in Applying Multivariate Methods in an Online Context (poster version)

Abstract:

The roots of modern multivariate analysis trace their roots back to the 1901 work of Karl Pearson. Despite the well documented benefits of using these techniques for offline analysis, the refining and petrochemical industries have been relatively slow in adopting the technology for online applications. This is surprising considering their enthusiastic and transformational adoption of more computationally intensive tools such as model predictive control.

In this presentation, we describe our experience in using multivariate methods to complement existing approaches and to create novel high value solutions.

480945 Hypothesis Investigation without a Designed Experiment

Abstract:

A designed experiment enables one to infer causal relationships between input variables and the variable of interest. When investigating a hypothesis related to a chemical process issue, a well-designed experiment can provide strong evidence to confirm or refute the claim. What about situations when a designed experiment is not feasible? This situation is common in a chemical manufacturing environment due to the high costs and potential safety concerns associated with recreating the process conditions that caused the original problem. One of Dow's plants faced this challenge as they sought to investigate the claim that the source (supplier) of a raw material was responsible for the lower concentration of active material in the final product. The raw material of interest is highly toxic and has to be manufactured on-site whenever possible to minimize its transport. Occasionally it is obtained from an off-site source when it cannot be produced in sufficient quantities on-site. A combination of visual methods and multivariate modeling techniques were used to examine historical plant data over a period of 3 years. The frequency of the data was hourly (averaged), and approximately three dozen variables were used from multiple sources: online process instruments, online plant analyzers, and the quality assurance lab. The hypothesis was convincingly refuted in the absence of a designed experiment.

480918 Determining Product Specification for Count Data

Abstract:

In order to ensure that a business can produce minimum amount of out of specification product, it is critical to include the statistical knowledge of its process and measurement capability in determining the product specification. This can be done by following "Big Data" concepts, utilizing historical process and/or lab data. As a standard six-sigma methodology, the capability of measurement is commonly evaluated from P/T ratios. However, the evaluation of P/T ignores within-product variation, and is inadequate for non-normal data distributions and/or concentration-dependent standard deviations. We propose a more generally applicable approach, in which we focus on confidence levels (or "critical fallout values") rather than on P/T ratios. Our approach allows for product variation and for all possible data distributions, including discrete distributions.

At The Dow Chemical Company, we have started to apply our generalized approach to the evaluation of method capability in the case of Poisson-distributed data. We will illustrate the merits of our approach from our experiences in this specific application.

480669 EMI Command Center

Abstract:

Dow, being one of the largest and long standing startups (118 years old), accrues immense chemical production process data. It's often said that a journey of a thousand miles starts with one step. This holds true for any effort to harness the power of the data in plant historians, laboratory information management system (LIMS), and online analyzers. Most of our data is underutilized and simply archived in databases and unattainable due to the manner in which data systems were designed and data was stored prior to the age of "Big Data". In order to extract the most significant and immediate value from that data, process design knowledge and operational experience can be used to summarize and aggregate plant data to expose the most relevant information in the right context. At Dow, we want the data to start working for us, rather than us having to work to get the data. The goal of Enterprise Manufacturing Intelligence (EMI) is to make these high value relevant data available in real time to all levels of production management (Operations to Plant and Business Management). Our approach in Manufacturing Analytics is to start small, look for small data opportunities within our plants and operations, and build momentum with success stories. In our quest for total data domination, we are already improving reliability, reducing costs, increasing value, and providing safer operations.

Building a command center for our EMI efforts would mean that we would have all our EMI implementations being monitored through one channel. A command center for EMI would bring the chemical industry to a completely new level of acuity with respect to operations.

480677 An Alternative Control Chart to Monitor Low Level Count Data

Abstract:

Control charts use statistics to help translate patterns in data into decisions. Data-to-Decisions is critical in CPI since early detection of deviations helps ensure safety, high quality, and productivity. Control charting is mainly applied to normally distributed data, but control charting techniques are available for tracking count results on failures, spills, equipment breakdowns, defects in polymers, punctured films, thin spots in yarn, etc. For these kinds of discrete data, the count control chart (c-chart) is the de facto industry standard.

The c-chart will perform well at high count rates, but can be slow to respond to process changes at low count rates. In failure-process monitoring, the time-between-events chart (t-chart) was developed as an alternative to the c-chart for low count rates. At The Dow Chemical Company we have extended the use of the t-chart to monitoring of defects in bulk chemical product. The t-chart allows for the detection of changes by the basic SPC rule, where the c-chart requires supplementary rules. More importantly, the time efficiency of the t-chart was found to be considerably better than the efficiency of the c-chart for detecting process changes in test data sets, especially for detecting process deterioration. Here the efficiency of the t-chart is several orders of magnitude better than the efficiency of the c-chart, greatly improving Data-to-Decision speed.

To paraphrase a statement from Leo Chiang in the Big Data feature in March 2016 edition of CEP magazine, 'Big Data Analytics expresses that the potential value is from the thoughtful collection of data combined with knowledge to answer complex questions.' If you have a large volume and/or velocity of data with low counts, the t-chart has potential to accelerate the detection of change relative to the c-chart, putting this Big Data Analytics concept into practice.

479593 Challenges and Benefits of Aligning and Reconciling Process Data for Seasonal Processing Industries

Abstract:

Industrial chemical manufacturing plants produce both chemical products and data, the latter in increasingly large amounts. While the former is obviously the raison d'être of the plant, it is becoming increasingly evident that the data both has intrinsic worth, and is often undervalued. The data originates from many sources such as pressure/flow/temperature transducers throughout the unit operations, records produced by analytical chemists in adjacent laboratories of day-old product, external environmental and/or financial conditions, and perhaps feedback from customers, partially filtered by the sales and marketing division. In many instances, all this heterogeneous data (at times of dubious quality) is stored in separate and incompatible databases behind imposing barriers. This makes it hard to draw any meaningful conclusions from this wealth of data. And this is just for one plant, what would happen if one wanted to compare across multiple sites?

Over twenty years ago, the advantages of fully exploiting captured data was recognised, and mathematical optimal strategies for producing a reconciled and consistent data set were proposed by a number of authors. Typically these strategies were cast as constrained optimisation problems, where the intent was to produce a set of adjusted process measurements that satisfied known mass and energy balance constraints. While academic interest seems to have waned somewhat, such approaches are still relevant, and plausible even with the large data sets routinely generated today. However simply constructing a reconciled data set is only a first step in what is potentially possible.

This work describes the construction of a reconciled and aligned database of three separate large-scale milk powder plants in New Zealand. The production of milk powder has some special challenges, compared to say bulk chemicals. First, given that the raw material (from cows) is a biological product of eating grass, in different geographical locations, there are many uncontrolled, and strongly correlated seasonal variables. Second, milk production in New Zealand is largely seasonal and typically only operates for 9 months of the year with an annual 3 month period used for sometimes significant plant upgrades or changes. Consequently, data captured in one season could be very different to the next, due to seemingly small plant changes. Finally, quantification of the end-properties of the milk powder are subtle, challenging to measure repeatably, and can be hard to relate to production conditions due to the complex multivariate nature of the underlying physical-chemical relations.

This motivation behind the creation of this database is to investigate the potential of 'Real Time Quality' in the production of Instant Whole Milk Powder (IWMP). Increasingly across the dairy industry, the focus of producers is shifting from maximising production to maximising quality, and higher-value milk powders and premium products, which necessarily also have higher requirements in terms of performance and composition. This covers a very broad range of attributes, including dozens of measurements each of physical, functional, sensory, and microbiological factors, not all of which are explicitly controlled at the time of manufacture. Functional properties, such as the dissolution behaviour, taste, or texture - given their inherent qualitative characteristics - are challenging to control and quantify accurately, and they are often not tested regularly. The determining physical causes are not always well known, and plant operators may rely on rules of thumb, or may not even have a chance to affect functional property outcomes if the test results are not timely. However it is advantageous that future quality control is performed in real time to prevent many tonnes of off-specification powder being produced before detection by the infrequent and delayed offline measurements.

It was hypothesised that multivariate regression utilising a very broad and deep dataset, containing many process and quality variables from several different powder plants producing the same premium product, across several years of production seasons, may give sufficiently varied input data to allow prediction of powder functional properties, suitable for real time decision making. This avoids the problem of insufficient excitation in the raw data from plants operating at steady-state to make meaningful models. This firstly required construction of a dataset, which was a vast undertaking to combine and align many sources of data, spread over time, geography, and changes in plant design and operating methods.

This approach distinguishes between three distinct types of measured data: ***X data***, which comprise of the standard process measurements such as temperature, pressure, and flow; ***Y data***, which comprise of at-line hourly measurements of in-process powder physical properties; and ***Z data***, which are the final powder functional properties. The X data are typical measurements that exist in any process plant, however Y data are key physical powder measurements, such as fat, protein, and moisture content. The Z data are typically prescribed by customers or regulatory bodies, and may be challenging to change or improve, however Y data may be expanded with newly introduced measurements, such as bulk density or particle size distribution, which it is hypothesised will improve prediction of functional properties by bridging the gap between measured process data and final powder properties.

These three data groups are stored in different databases, and not always with clear methods to cross-reference them. A further complicating issue is that the powder transport chain also creates a significant challenge where the actual production time of any ***Z-data*** quality sample taken from packed powder bags cannot be confidently known, due to holdup time and mixing of powder from large blending bins and silos.

Construction of the database required close examination of the specific operating and sampling processes at each plant, and their associated data storage methods. This uncovered a large number of special cases in the data streams, which must be detected and cleaned up in every case, which requires time consuming manual programming, and is not easily generalised to other plants. This is a fundamental feature of many types of industrial data however, and the cleaning up of data idiosyncrasies is a key requirement to create a useful dataset, which would greatly benefit from development of advanced or smart data processing methods. It is also, in our experience, a key reason why properly aligned data is so rarely constructed.

Missing data is a further key challenge to surmount, as the routine plant cleaning cycles cause a significant fraction of plant data to be out of range at all times. For example in a plant with four parallel milk evaporator trains, one or more are always out of service for cleaning, and the process temperatures recorded are either the ambient temperature, or that of the cleaning fluids, and must be removed from the data set when comparing process conditions. Missing data can be imputed, which may have an effect on the behaviours observed, or alternatively samples with any missing data in their row can be dropped. However the dropping of entire rows causes severe data reduction at times, up to 100% of the data in plants with parallel unit operations where one is always offline, and clearly such an approach is untenable. Consequently other methods have been proposed such as creating composite variables from parallel unit operations, with some success.

Carefully framing the questions to be asked of the data is a key factor in preventing unnecessary data reduction, by informing the optimal data arrangement or imputation. This planning was also vital in selecting which parts of the data to use for predictive modelling. Changes in plant equipment or operating methods cause structural changes in the data set and change or break underlying relationships. For example, the full data set can be used to find similarities and differences between the three plants producing the same high-value product, however there are significant differences in plant design between them, and they do not have the same relationships between process conditions and product quality. Product quality is much better investigated when using data from only one plant at a time, and produces models which may be used to inform process conditions particular to that plant.

While significant work has been undertaken in prediction of functional properties, there is also great value produced solely by acquiring and organising this dataset, which had not been previously undertaken due to the difficulty and effort required. Visualisation of this plant data using novel methods or arrangements can elucidate many behaviours, differences and similarities between the different plants, products, and times of the season. Data visualisation is a rich subject with great value, which is rarely in our experience ever well explored or exploited by plant process control engineers. In order to generate information from rich visualisation however, of course the appropriate dataset needs to be collected and aligned, which may be a prohibitive undertaking without careful planning, design and intelligent processing methods.

The paper will illustrate uses of this data set, showing visualisation strategies for time-varying multivariate models, and how plant engineers can quickly assess the current plant status compared to past runs, as well as work in finding predictive models for use in Real Time Quality production.

480355 Big Data Approach to Fault Detection and Diagnosis in Batch Processes Using Nonlinear SVM-Based Feature Selection

Abstract:

Simultaneous achievement of high process efficiency, safety, and profitability is of utmost importance in modern process industries [1]. A major challenge in industrial applications is the rapid detection and identification of process faults in order to sustain a safe operation and minimize losses in productivity [2]. Statistical process monitoring (SPM) algorithms [3] are often used to detect the deviation from normal operating regimes. A growing number of studies have focused on data-driven process monitoring by using the Tennessee Eastman [4] and Pensim [5] benchmark datasets for continuous and batch processes respectively [6-9].

Batch reactor processes are widely used in chemicals, food, and pharmaceutical industry. These processes involve a considerable number of interconnected variables. In addition to inherent non-stationarity, batch processes are characterized with finite duration, nonlinear response, and batch-to-batch variability [10-12]. High complexity as well as dimensionality of batch processes impose a big challenge in fault diagnosis. Most novel techniques for fault detection and identification have focused on continuous processes, and the need of monitoring algorithm development for batch processes is evident [13].

We present a new data-driven framework for process monitoring and intervention in batch processes. Central to the framework are novel theoretical and algorithmic developments in support vector machine-based dimensionality reduction which improve accuracy, guide fault diagnosis, and encapsulate highly nonlinear relationships. We will discuss critical data processing and feature extraction steps specific to batch processing. Our methods will be applied to a recent extensive benchmark dataset [13] which features data describing 90,400 batches with numerous and diverse fault types. The analysis framework aims for early detection of faulty batches and enables intervention to reduce loss of profit.

References

1. Floudas, C. A., Niziolek, A. M., Onel, O., & Matthews, L. R. (2016). Multi Scale Systems Engineering for Energy and the Environment: Challenges and Opportunities. AIChE Journal.

2. Subramanian, S., Ghouse, F., & Natarajan, P. (2014). Fault diagnosis of batch reactor using machine learning methods. *Modelling and Simulation in Engineering*, *2014*, 15.

3. Joe Qin, S. (2003). Statistical process monitoring: basics and beyond.*Journal of chemometrics*, *17*(8 9), 480-502.

4. Downs, J. J., & Vogel, E. F. (1993). A plant-wide industrial process control problem. *Computers & chemical engineering*, *17*(3), 245-255.

5. Birol, G., Ündey, C., & Cinar, A. (2002). A modular simulation package for fed-batch fermentation: penicillin production. *Computers & Chemical Engineering*, *26*(11), 1553-1565.

6. Van den Kerkhof, P., Gins, G., Vanlaer, J., & Van Impe, J. F. (2012). Dynamic model-based fault diagnosis for (bio) chemical batch processes. *Computers & Chemical Engineering*, *40*, 12-21.

7. Wuyts, S., Gins, G., Van den Kerkhof, P., & Van Impe, J. (2015). Fault identification in batch processes using process data or contribution plots: A comparative study. *IFAC-PapersOnLine*, *48*(8), 1282-1287.

8. Chiang, L. H., Braatz, R. D., & Russell, E. L. (2001). *Fault Detection and Diagnosis in Industrial Systems*. Springer Science & Business Media.

9. Mahadevan, S., & Shah, S. L. (2009). Fault detection and diagnosis in process data using one-class support vector machines. *Journal of Process Control*, *19*(10), 1627-1639.

10. Dahl, K. S., Piovoso, M. J., & Kosanovich, K. A. (1999). Translating third-order data analysis methods to chemical batch processes. *Chemometrics and intelligent laboratory systems*, *46*(2), 161-180.

11. Smilde, A. K. (2001). Comments on three way analyses used for batch process data. *Journal of Chemometrics*, *15*(1), 19-27.

12. van Sprang, E. N., Ramaker, H. J., Westerhuis, J. A., Gurden, S. P., & Smilde, A. K. (2002). Critical evaluation of approaches for on-line batch process monitoring. *Chemical Engineering Science*, *57*(18), 3979-3991.

13. Van Impe, J., & Gins, G. (2015). An extensive reference dataset for fault detection and identification in batch processes. *Chemometrics and Intelligent Laboratory Systems*, *148*, 20-31.

480990 Industrial Application of Multivariate Modeling Methods for Process Forecasting and Model Predictive Control

Abstract:

Multivariate (MV) methods are commonly used for characterization and statistical process control (SPC) monitoring of batch processes. Extensions of these modeling methods can also be used for predictive purposes to estimate future trajectories of important process parameters or final conditions of batch type processes. This talk provides a brief overview of the multivariate forecasting methods, referred to as imputation in literature, as well as the introduction of a new regression based method. Application of these imputation methods within an optimization framework is provided to demonstrate their use for process optimization and model predictive control (MPC). This is particularly useful in batch processes where traditional MPC technologies are not suitable.

480913 Practical Challenges and Opportunities in Applying Multivariate Methods in an Online Context

Abstract:

The roots of modern multivariate analysis trace their roots back to the 1901 work of Karl Pearson. Despite the well documented benefits of using these techniques for offline analysis, the refining and petrochemical industries have been relatively slow in adopting the technology for online applications. This is surprising considering their enthusiastic and transformational adoption of more computationally intensive tools such as model predictive control.

In this presentation, we describe our experience in using multivariate methods to complement existing approaches and to create novel high value solutions.

13TH GLOBAL CONGRESS ON PROCESS SAFETY

The Capability Engine, Analyzing the Past to Protect the Future

Lawrence Short
DuPont Sustainable Solutions
500 Studio Drive, Virginia Beach, Virginia
Lawrence.S.Short@dupont.com

Alfonsius Ariawan
DuPont Sustainable Solutions
1919 Minnesota Court, Mississauga, Ontario
Alfonsius.B.Ariawan@dupont.com

Prepared for Presentation at
American Institute of Chemical Engineers
2017 Spring Meeting and 13th Global Congress on Process Safety
San Antonio, TX
March 26 – 29, 2017

The Capability Engine, Analyzing the Past to Protect the Future

Lawrence Short
DuPont Sustainable Solutions
500 Studio Drive, Virginia Beach, Virginia
Lawrence.S.Short@dupont.com

Alfonsius Ariawan
DuPont Sustainable Solutions
1919 Minnesota Court, Mississauga, Ontario
Alfonsius.B.Ariawan@dupont.com

Keywords: Learning, Capability Building

Abstract

Hazardous operational environments require the ability to identify and minimize process and personal risk variables. In Northern Alberta, this is challenging, with many upstream oil companies sharing a large common population of contract workers. The Personal Safety Collaboration brought owners together to maximize the effects of a comprehensive learning model. Through this initiative, the Capability Engine was conceived and designed. Using theoretically sound best practices, descriptive assessment data is examined with our analytics model combined with field based performance and environmental data. This combination renders an inferential and predictive tool enabling prescriptive action while alerting of the historical leading indicators of incidents.

1. Introduction

Through our work with a client in Northern Alberta, we have developed a Learning Effectiveness model to enhance operators' learning journey. We did this by incorporating more structures and learning assurances throughout an operator's learning cycle. The model is depicted in Figure 1 and is explained elsewhere in more details[1].

An important element of our model is the analysis of learning measures, specifically the Kirkpatrick K3 measures, that link field performance to training variables. The K3 measures are made up of field performance indicators, such as those related to operational discipline, audit performance, incidents and near-misses, as related to specific tasks (It is differentiated from K1 and K2 measures which focus on the perceived effectiveness of training tools and delivery, and the understanding of knowledge through testing and field demonstrations, respectively. The reader is referred to another paper by the same authors for further explanation[1]). Our instructional designers combine both the practical experience that we have obtained through client work and instructional theories to design the various assessment protocols in order to determine the right Kirkpatrick measures. This allows us to link the resulting data to the various critical points of the operator's learning journey. With some basic analysis, we can then determine where failure modes have happened in the journey. For example, we are able to determine if the way the course content is delivered to operators lacks impact, or if there is a lack of supervisory reinforcement in the field. This then allows the organization to address issues to further enhance learning vs. investing in more training session.

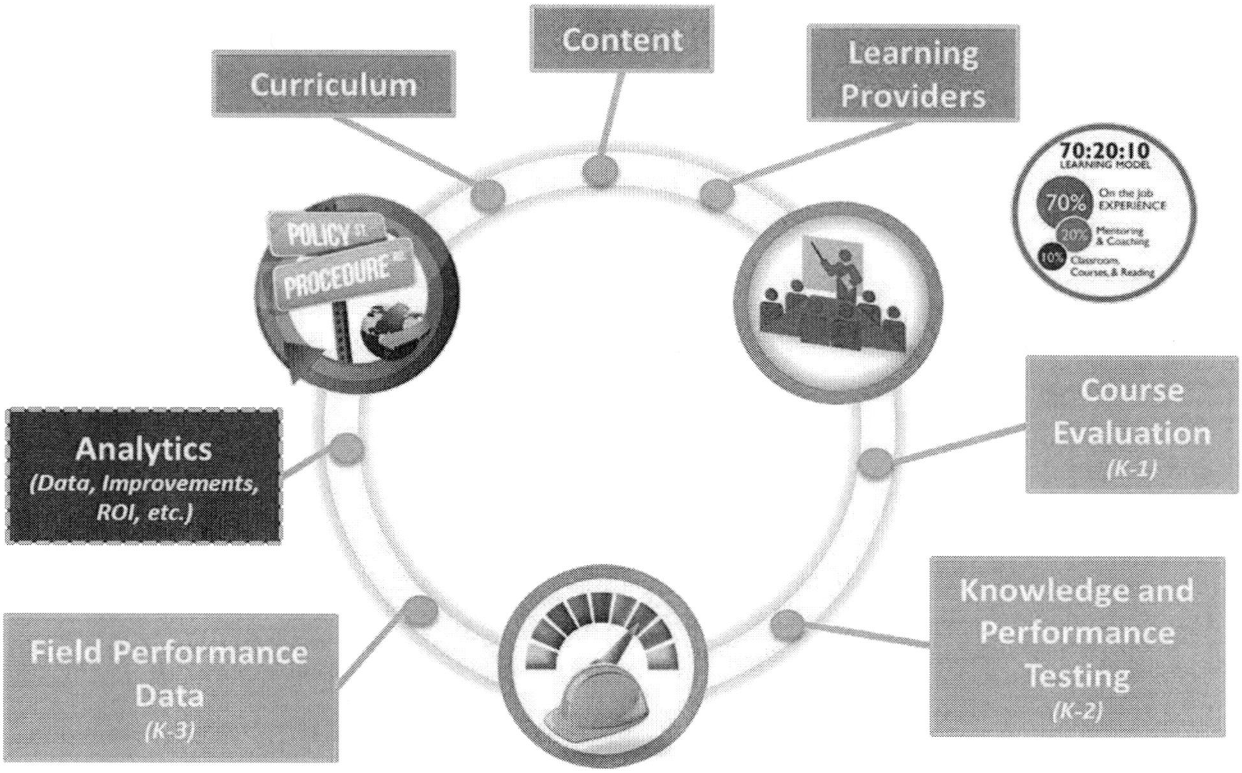

Figure 1. Learning Effectiveness model

In more recent applications of this model, we have also incorporated environmental data to the analysis. Our premise is that environmental factors may influence how operators learn, and create a bias in their field performance.

2. The Capability Engine

The capability engine of the learning effectiveness model refers to the curriculum, content and delivery mechanisms that facilitate knowledge transfer. These are continuously improved based on K1-K3 assessment results and the subsequent data analysis. For example, recall that K1 data tells us about the perceived effectiveness of training and its delivery. It is based on participants' rating of the facilities, instructor, quality of content, applicability of work, etc., complemented with our own measures of the same. K2 data, on the other hand, measures the transfer of knowledge through testing and field demonstrations. Lackluster K1 and K2 data, but an acceptable K3 outcome or performance suggest that learners do not think much of the training session and do not necessarily obtain the knowledge that is intended to be transferred. However, learners are able to practice in the field what are expected of them, and do so in an acceptable consistency. This is certainly a possibility. For example, this could be due to effective field supervision and strong peer support/coaching. However, this begs the question on the return of investment of the training sessions, i.e. what is the value of the training sessions? Or, more positively, one may want to ask: if we were to improve the quality of the training sessions, would we be able to further improve performance? If so, how?

On the flip side, if K1 and K2 measures are deemed to be acceptable, but the organization does not see the expected outcome or performance in terms of improvement in incident rate, near-misses, audit infractions, etc., then different questions may need to be asked. The issue could be in the conversion of training materials to practical application. Have learning objectives been defined correctly? Can we assure the fidelity of the course contents to these learning objectives? There may also be field variables or external factors at play that need to be addressed.

Before drawing any conclusion, the design and quality of the assessment questions/protocols (through which K1 – K3 data are measured) must also be analyzed. Poorly designed questions can result in data bias and/or in results that are not sensitive enough to input variables. These questions need to be continually checked and upgraded as new learnings are obtained, possibly complemented with greater number of input variables, such as environmental variables, to strengthen correlation analysis (big data analysis). In designing these assessment questions/protocols, we also consider the three domains of learning to ensure comprehensiveness:

- **Psychomotor:** The ability to perform tasks by manipulating objects and / or moving one's body in space to complete a task
- **Affective:** The internalization of attitudes, beliefs that are best measured through observation of behavior
- **Cognitive:** knowledge, comprehension, application, analysis, synthesis, evaluation / judgment

3. A New Way of Learning

One of the more significant environmental variables that influences learning performance is the demographic shift in workforce. The rapid replacement of operators with younger personnel has challenged our traditional method of knowledge transfer, primarily due to the way the younger generations are learning. Analysis of K1 – K3 data forces us to rethink how we should design and deliver learning contents. The reality is that technology advancement has changed the way information is transmitted and therefore, consumed. The ease with which information is transmitted often put us in a constant state of distraction and make us the victim of information overload. In response to this new reality, we adjust. We now tend to consume information in small 'chunks'. We do not need to look far for evidence: YouTube, Tweeter, Ted Talks and Instagram are just a few examples of platform that transfer information in short bursts to their intended audience – effectively. So, why do we still expose our operators to days-long training sessions, where information typically flows one way – from the instructor to the participants? Analysis of K1 – K3 data suggests that micro-learnings are also more effective in knowledge transfer in operational setting. The use of multi-medias and interactive exercises have also been found to enhance learning, especially if they are designed and delivered to affect the learners' emotions.

5. Summary

In this paper, we describe a learning model that is designed to be continuously improving through the analysis of K1 – K3 measures. These measures need to be thoughtfully designed in order to ensure comprehensiveness, sensitivity to input variables and soundness in their cause-effect relationships to the various elements of the capability engine. The results of the analysis allow us to pin point and address critical failures in our operators' learning journey. A finding from our work with clients in this space suggests the need to focus on micro-courses as a possibly preferred mechanism to transfer knowledge, with the use of multi-media and interactive exercises. Such approach anecdotally leads to better knowledge retention, especially if designed and delivered to reach the learners' emotional level.

6. References

[1] American Institute of Chemical Engineers (2017). *Ensuring Learning (Not Just Training) – Measuring the Impact of Learning on Business Performance*. New York, NY: Lawrence Short, Alfonsius Ariawan

479452 "Getting More Out of Your PHAs with Data Analytics"

Abstract:

Process Hazard Analysis (PHA) studies are required to identify process and operations risks for process plants. PHAs enable management and process safety staff to make decisions that improve safety performance, minimize downtime, and optimize utilization.

Operating companies also must make these key decisions under financial capital constraints and ensure maximum return on investment for safety. These decisions require a clear understanding of the interconnectivity of the risks associated with a facility's operations which are captured in a PHA. Most PHA studies generate recommendations based on residual risk and consequence levels, but they do not directly address how safeguards and recommendations are linked to the hazard scenarios. By improving the way PHA data are analyzed, using data analytics, additional insight is generated that is typically hidden, or not considered, when traditional PHA approaches are followed. These insights include more intuitive identification of critical safeguards (barriers), better prioritization of recommendations from a cumulative risk perspective, and visual aids to improve awareness and communication of risk drivers to operational staff and decision makers.

Traditional PHAs require a substantial investment in time and resources, which results in a potential gold mine of useful data that can be further extracted and exploited. With an ever shifting focus on new technology and innovation, data analytics is the next key step for improving risk management practices within organizations and the industry as a whole. Taking PHA data analytics further, "Big Data" data mining techniques can be utilized to extract information from the collected data and expose trends and patterns in industry PHA data. These trends can be benchmarked across various industries providing an advanced level of decision making insight not available from a single PHA. Statistical analysis of the trends and deviations, along with analysis of incident data can also predict increasing levels of risk, thus creating strong leading indicators to aid in decision making before incidents occur. This paper showcases a case study for a unit within a refinery which demonstrates how application of PHA with data analytics can greatly improve decision making.

479294 Prevent the Next Catastrophe

Abstract:

With the increasing complex manufacturing processes and environmental regulations, and the challenges to retaining knowledge from retiring employees and transferring it to the newer generation, operations managers face multiple challenges such as staying competitive, insuring safety, maximizing production, and reducing operational risks and breakdowns. In the world of legacy, traditional manufacturing practices rely on repetitive procedures, overwhelming human interventions leading to costly errors. The fourth industrial revolution presents an unprecedented mind shift in Manufacturing Operations Management with the emergence of Cyber-Physical Production Systems, Augmented Reality Technologies, Big Data Analytics and Artificial Intelligence. New practices consist in creating systems that learn from data and deploy the extracted knowledge in real time so the end-user can have a single version of the truth, rather than abstract mathematical certainties. Thus, sudden equipment failures and process disturbance are being predicted, and costly damages are prevented. Integration Objects provides an Industrial Internet of Things platform that allows end-users to remotely and securely assess and monitor the reliability of their assets while maximizing profitability. Our innovative approach uses data driven hybrid models combined with expert rules and fault propagation models that can be deployed online for real time execution. Our IIoT platform is an integrated, flexible, scalable and user friendly graphical environment that allows the deployment of industrial scale applications.

In order to address the market shortage of resources, Integration Objects has made the technology easy-to-use, affordable, and accessible to the average user for a quick deployment. It allows increasing safety by avoiding false warnings, and boost gains by seizing any optimization opportunity.

480602 On Mapping the Operating Window of Processes

Abstract:

When you have a new process on your hands, be it a lab scale or a large scale manufacturing process, how do you map your process operating window?

Theoretical modeling can only take you so far, and the answer it provides can be too generic for your rather specific case. A good deal of empiricism seems to be generally applied when it comes to mapping the operating window.

Practitioners typically espouse educated trial-and-error in order to search for upper and lower levels of the process variables which can render a fail-safe outcome: the so called operating range of these variables. More often than not, a restricted range is established for the lack of a systematic search procedure. The amount of time necessary to conduct a full search of the operating range of each variable may be easily prohibitive under trial-and-error. When operating ranges are restricted, so are the possibilities of process optimization.

In the theory of design of experiments, Screening designs are the very first line of attack to process improvement. In order to screen variables for their relative importance, Screening designs place experimental runs at the (high/low) edges of the operating ranges. It is *sine qua non* that these experimental runs be fail-safe so that model parameters can be properly estimated with a few runs. Hence, prevention of fail runs is the motivation for restricting the operating window, i.e. making it narrower than it could be, so as to be able to implement Screening designs smoothly.

As such, the establishment of the operating window is a pre-condition for the optimization of new processes. If you do not know the high/low levels, you cannot screen variables. And if the operating window is established in a restricted form, it reduces the level of optimization that can be attained if some sweet spot within the true (and larger) operating window is left out.

The work presented here offers a systematic method for mapping the operating window with minimal restriction. This method makes use of Sequential Design of Experiments. It starts out with an ordinary Screening design and, as fail runs surface, it makes use of Logistic regression to establish the initial location of the boundary of the operating window within the multivariate space of process input variables. Iteratively, new runs are placed at the boundary, new fail runs surface, and the boundary is relocated after re-estimating the Logistic model with new information. A heuristic is used to induce convergence to the true boundary after a few iterations.

With this method, experimental runs are employed with the sole objective of optimally learning the location of the operating window, which is in average a more cost-effective manner of mapping vs. trial-and error. This, in turn, opens up the possibility of mapping the operating window with minor restrictions within a manageable timeline.

An additional advantage of this method is that the fail-safe runs it generates form, in some cases with the need for augmentation, an experimental design which can be used for estimating Screening models altogether.

479952 Utilizing Big Data for Development of Soft Sensor

Abstract:

Conventional static soft sensor does not handle the dynamic of processes adequately. Instead the sequential historical data should be considered when developing the soft sensor. It follows that the resultant soft sensor is able to extract dynamic information. With advancement of data storage facilities, there is a deluge of data. Excessive computation burden can result if the complete big data are used. There is thus a need to utilize the data diligently. To this end, a latent variable model (LVM) based strategy to select useful data to overcome the problem of variable correlations and large number of sample is proposed. The uncertainty information obtained from the Gaussian process (GP) model is used as the selection criterion. The combination of LVM and GP model is developed together with an update scheme for advance improvement. This is done in order for the soft sensor to better reflect the current status of the process. The proposed method is applied to the data from an industrial process.

479943 A Multi-Tier Approach for Steam Cracker Revenue Optimization

Abstract:

Optimization models to maximize profits in steam crackers have been implemented widely in the industry. A complete plant simulation including all furnaces, hydrogenation reactors and recovery section results in a highly non-linear problem. Traditional optimization approaches utilize either simplified linear programming models or a steady state plant model in order to reduce the computational costs.

Typically, the optimal process conditions valid for the start of run do not remain optimal for the end of run, owing to the coke deposition, which impact fuel requirements and ultimate plant yield. In order to achieve a real-optimization, it is hence imperative to not only consider the energy costs incurred in the furnace and the compressors, but also to estimate the impact of the dynamic nature of the coke deposition.

This paper describes a novel multi-tier approach [Figure 1] to decompose the dynamic nature of the non-linear problem of steam crackers optimization. The furnaces are simulated rigorously using SABIC in-house cracking kinetics model . At each time-step, the coke-thickness deposited on the cracking tubes is estimated utilizing the rate of coking reaction computed at the previous time-step.

A pseudo-steady state optimization problem is then formulated which is solved in two-tiers.

- In the inner loop, process conditions are generated by optimizing the yield of a simplified recovery section model integrated with the furnace model (*Yield Optimizer*).
- In the outer loop, the profit is computed using a rigorous recovery section model integrated with the furnace model. Based on the profit difference between two consequent iterations, constraints on the process conditions are relaxed in order to run the yield optimizer recursively.

In this study, our two-tier and dynamic optimization approach is demonstrated via practical example to improve the computational efficiency and to achieve realistic profit-based optimization of olefins plant.

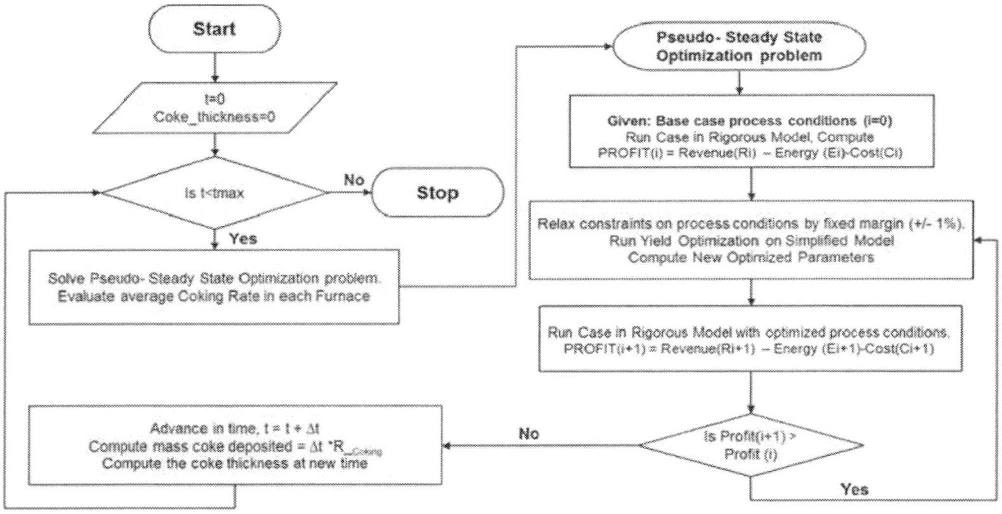

480736 Making Cognitive Solutions Real Today -Stories and Examples

Abstract:

IBM and others have aggressively invested in the advancement of Cognitive technologies such as Natural Language Processing, Machine Learning and Image Recognition, over the past decade . Cognitive driven solution are now having a profound impact on social and economic systems around the world. In this session, IBM will walk the audience through relevant use case and describe what's involved in designing, building and implementing cognitive solutions in a chemical engineering environment. The session will conclude a live demonstration of cognitive solutions.

TERADATA

Leveraging Advanced Analytics to Extract Business Value from Internet of Things Investments

Johnny Gipson, Senior Consultant, Oil & Gas Practice
Waseem Shaik, Practice Lead, IoT Analytics

Agenda

- Introduction
- Industry Challenges
- Teradata Approach
- Teradata Solutions
- Teradata differentiators

2 TERADATA

IoT – What's New?

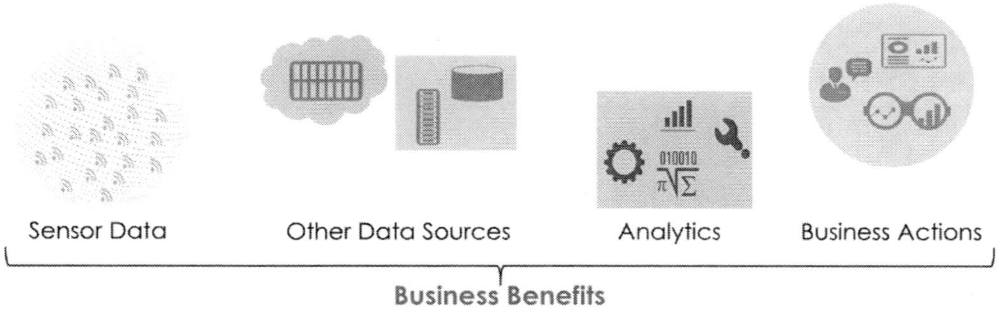

Right data, right insights, right actions driving business outcomes – revenue growth, customer satisfaction, operational efficiency

3 TERADATA.

What are Industry Leaders Saying

- **35%** of US manufacturers are currently collecting and using data generated by smart sensors to enhance manufacturing/operating processes
- Organizations capture but face challenge using data. **48%** spend the most time reviewing data for quality and consistency

Business Focus for IoT initiatives

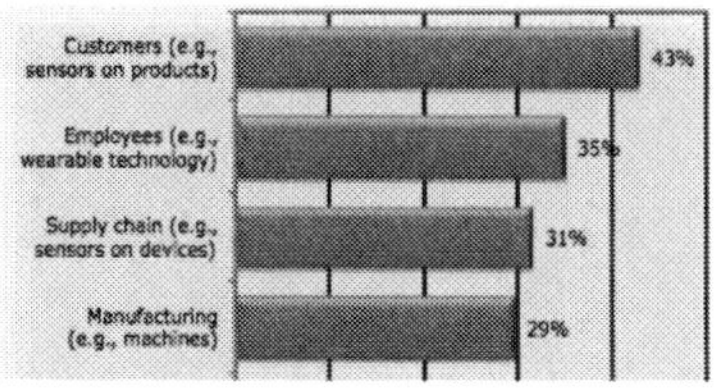

Issues with Implementing Technology

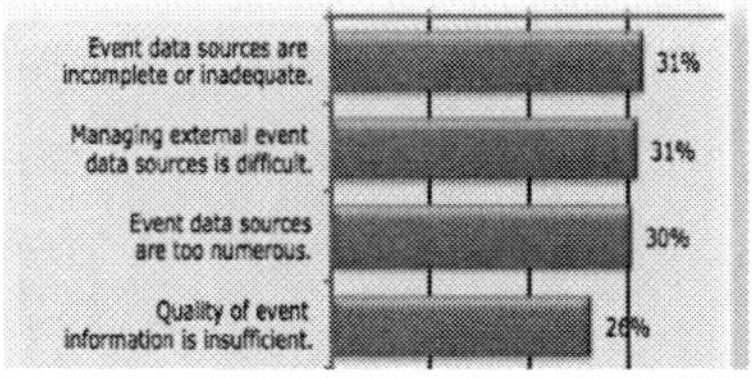

4 TERADATA.

Today's Challenges – Data Issues

It comes from		It can contain...		It has impact
Fleets: from lots and lots of similar things		**Outliers:** Which of my things are behaving differently?	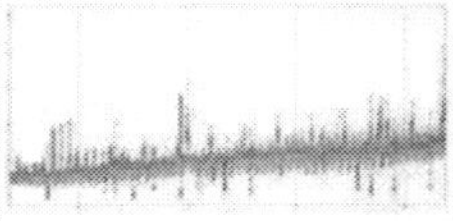	"Fleet-wide" 24/7 for holistic management
Systems: across the same big "thing" or facility		**Emergent behaviour:** Is my system changing to a new state?	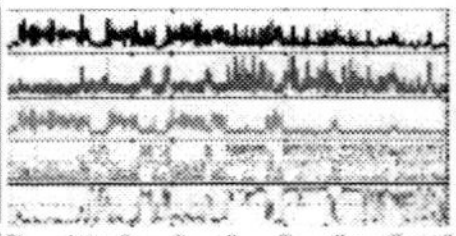	High-level KPIs at business units and facilities level
Collectors: "big models" or monitoring		**Events:** are there hidden signals?	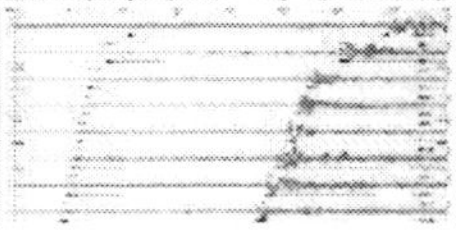	Performed at sub-second level and data kept for decades

5 TERADATA

Today's Challenges – Analytics Complexity

Mean Shift
Neural Networks
Affinity Analysis
Spectral Clustering
Feature Selection
Random Forest
nPath Analysis
Association Analysis
Nearest Neighbors
Support Vector Machines
Time Series Analysis
Geo-Spatial Analysis
Machine Learning
Clustering
K-Means
Regression
Dimensionality Reduction
Decision Trees
Scoring Models
Graph Analytics
Text Analytics

- Analytical Tools
- Model Generation
- Validity
- Accuracy
- Speed
- Scalability
- Repeatability

6 © 2016 Teradata TERADATA

Today's Challenges - Business link – Operationalizing of Analytics

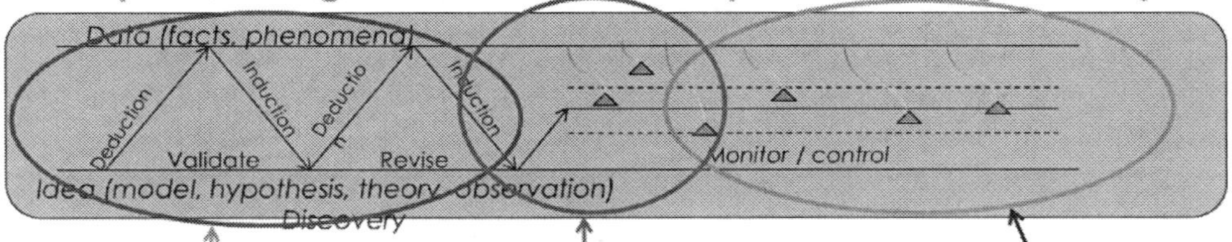

Engineering Focus
- Selecting right tool based on business need
- Setting up a scalable and repeatable environment
- Difficulty in converting business need into an engineering requirement

Implementation
- Missing: Taking discovered relations into production – quickly, efficiently, accurately
- Integration with BI tools, Engineering tools, Reports
- Platform decisions

- Business Contextualization
- Deriving meaningful business actions
- Adaptability of business
- Ability to monitor and measure the impact

7

TERADATA.

In Summary, Your Business problem looks like

The Problem

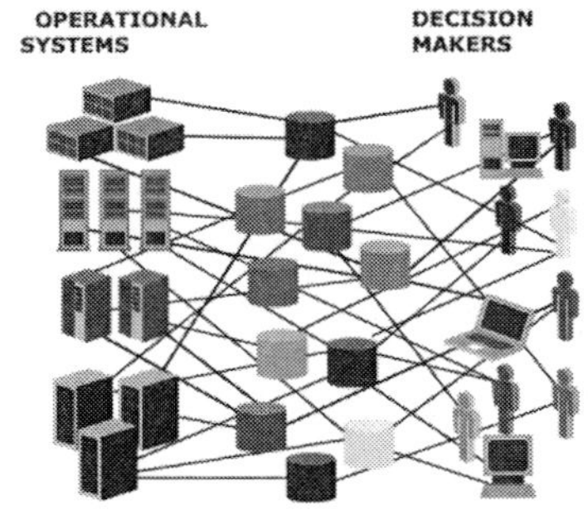

Proliferation of data marts resulted in fragmented data, higher costs, poor decisions

- Represent one silo of the business
- Bringing many silos of data can prove to be difficult
- Cannot keep pace with frequent changes in market conditions
- Decision by leadership is rarely made with complete data picture
- Creates poor communication in value chain
- Data Latency is technology or work process

8

TERADATA.

How can we provide decisions at the pace of business

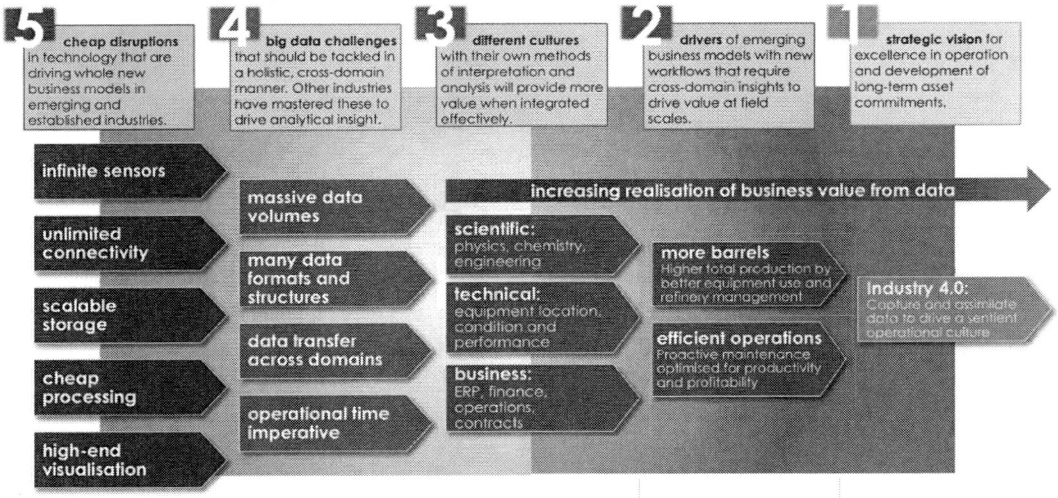

TERADATA

Teradata Approach & Solutions for Industrial IoT

AoT Approach – Overview

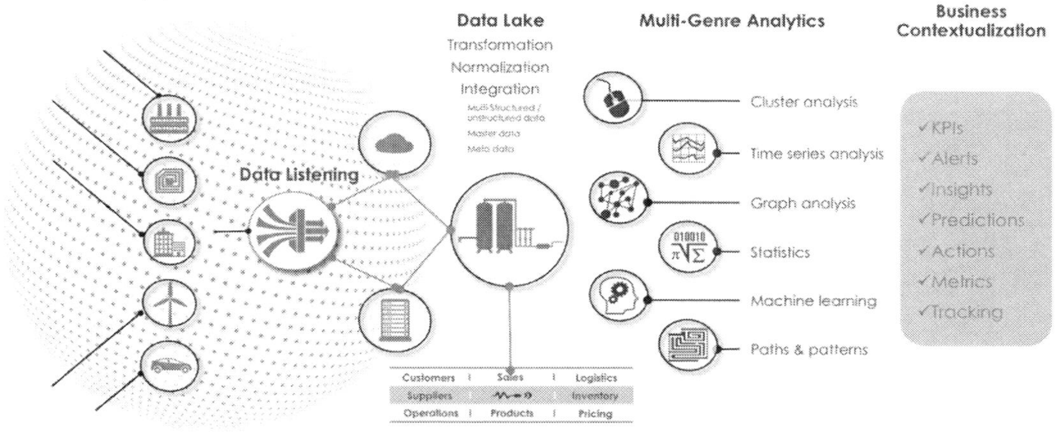

11 TERADATA

Oil & Gas Applications

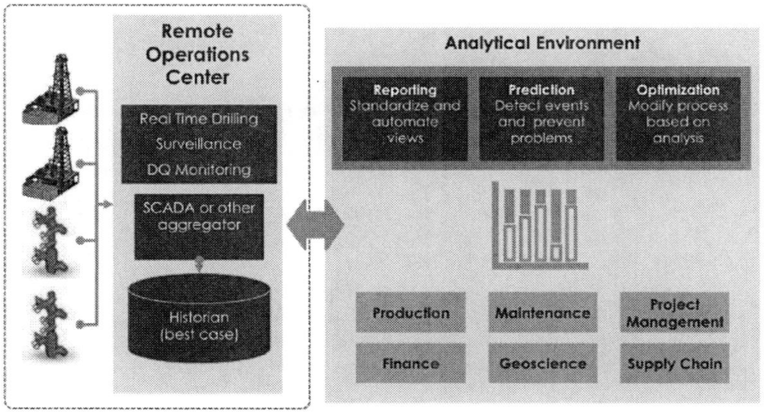

12 TERADATA

Contents

- Sensor Data Qualification
- Connected Factory
- Smart Assets
- Connected Supply Chain
- Services 360

13

TERADATA.

Solution: Sensor Data Qualification

Data Ingestion	Data Transformation & Storage	Data Analytics	Business Contextualization
DQ Stage 1	**DQ Stage 2**	**DQ Stage 3**	**DQ Stage 4**
Qualify Data Ingestion	Qualify Data Transformation	Qualify Data Analytics	Qualify Business Relevance

* DQ: Data Qualification

14

TERADATA.

Contents

15 TERADATA

Solution: Connected Factory

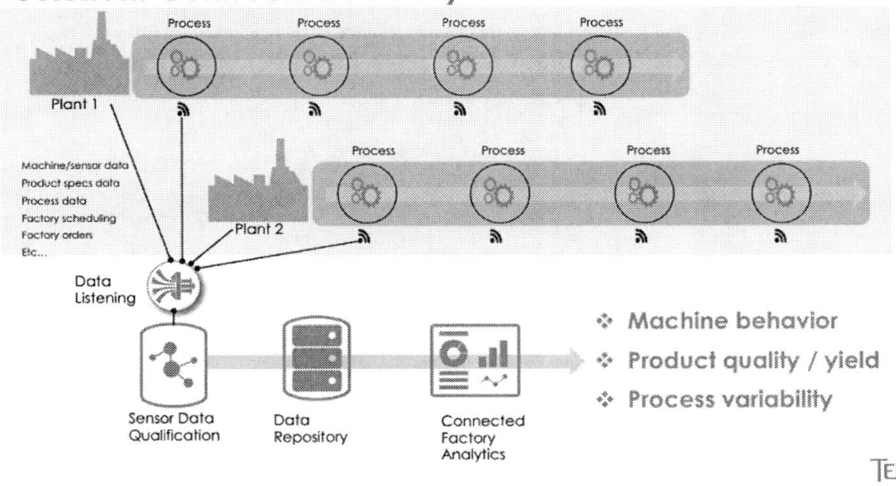

16 TERADATA

Connected Factory – Machine Behavior

Monitor and analyze machine logs and machine sensors as time series to predict machine failures and prevent unplanned downtime

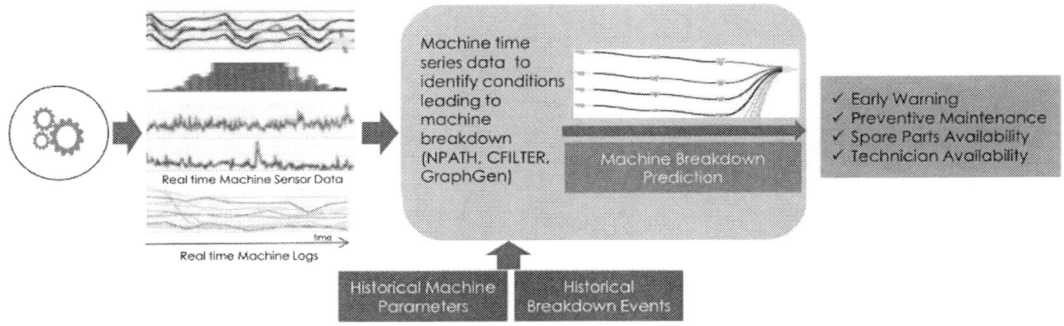

17

TERADATA

Connected Factory – Product Quality / Yield

Analyze product quality issues to identify conditions that lead to quality defects, yield issues, and build predictive models for detecting these issues

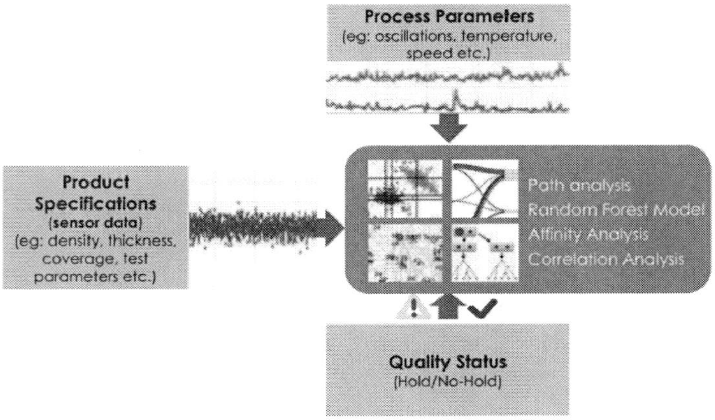

- ✓ Early detection of potential product quality issues
- ✓ Proactive adjustment of conditions that could lead to quality issues, or reduce yield
- ✓ Simulate yield / quality given a set of conditions (design of experiments)

18

TERADATA

Connected Factory – Process Variability

Analyze process variability impact to address process maturity issues, predict production yields/outputs, and reduce order commitment risk

Sub-Process Variability
(Process flow rules, Real time processing times)

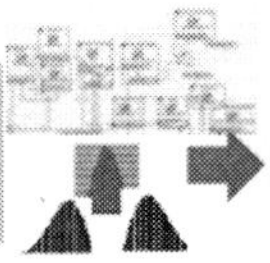

- Variability impact analysis
- What-if analysis & scenario modeling
- Alternate production path analysis
- Order risk analysis
- Raw material availability

✓ Increased productivity
✓ Faster time to process maturity
✓ Fulfillment predictions
✓ Risky order alerts
✓ Lowered work-in-process inventory
✓ Improved material availability

19

TERADATA

Contents

- Sensor Data Qualification
- Connected Factory
- **Smart Assets**
- Connected Supply Chain
- Services 360

20

TERADATA

Solution: Smart Assets

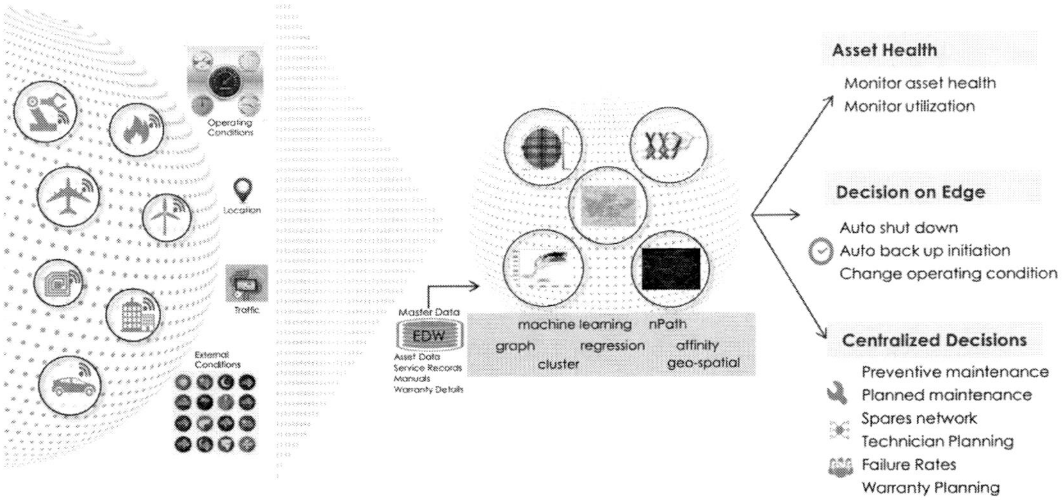

Smart Assets – Use Cases

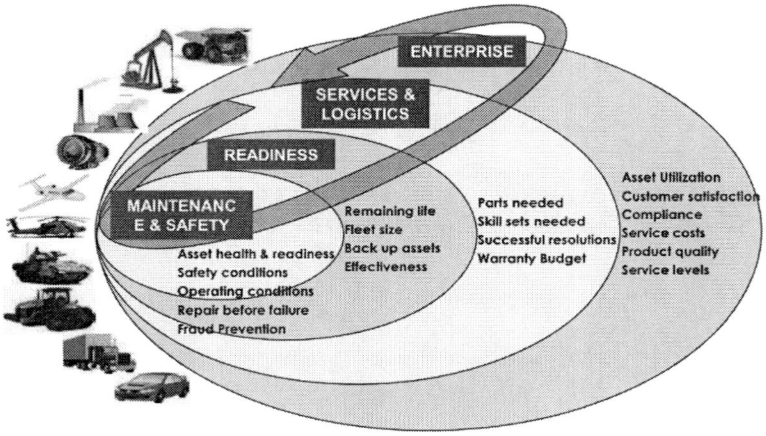

22

TERADATA.

Contents

23

TERADATA

Solution: Connected Supply Chain

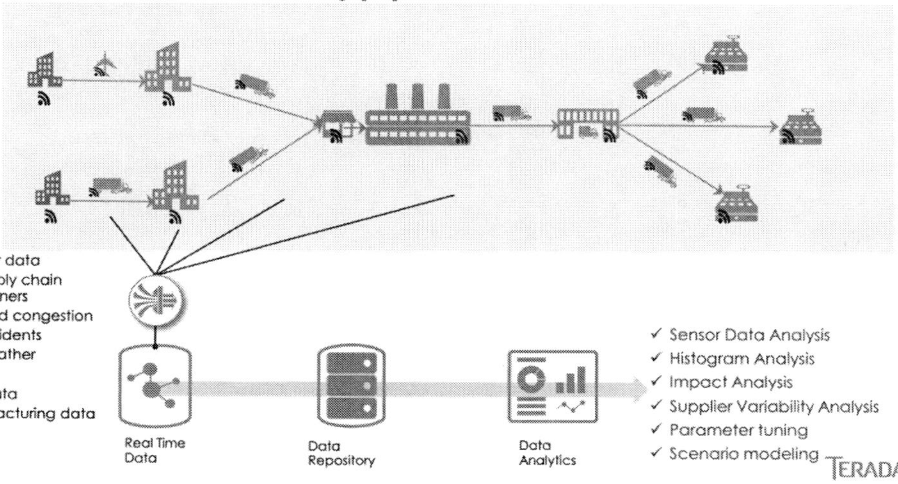

Partner data
- Supply chain partners
- Road congestion
- Accidents
- Weather

GPS data
Manufacturing data

- ✓ Sensor Data Analysis
- ✓ Histogram Analysis
- ✓ Impact Analysis
- ✓ Supplier Variability Analysis
- ✓ Parameter tuning
- ✓ Scenario modeling

24

TERADATA

Connected Supply Chain Maturity Map

Data
Enterprise Data (ERP & Transaction systems)

Operation
Reactive mode
History based
- Forecast
- Failure patterns (MTBF)
- Variability

Result
Target stock levels (SS/ROP)
Production schedules
Distribution/Logistics schedules
Changes happen after the fact (such as delays in receipts, delays in shipping times etc.)

Traditional Supply Chain Planning

Internal data systems effectiveness

Data
Partner Data Sources (suppliers, 3PLs)

Operation
Instant reaction mode
Reacts to instant changes (order not shipped, shipment progress not as expected etc.)

Connected Supply Chain Planning – Proactive

Collaboration across partners

Result
Expedites
Change orders
New orders
Stock gap alerts
Production change alerts
Demand change alerts

Data
Real time data
(Sensors, GPS, Social Media, market patterns etc.)

Connected Supply Chain Planning – Predictive

Real time data and collaboration

Operation
Predictive mode
Predicts based on real time data analytics

Result
Proactive changes to schedules, orders etc.
Automated decisions
Edge decisions
Demand shaping

25 TERADATA

Contents

- Sensor Data Qualification
- Connected Factory
- Smart Assets
- Connected Supply Chain
- Services 360

26 TERADATA

Solution: Service 360

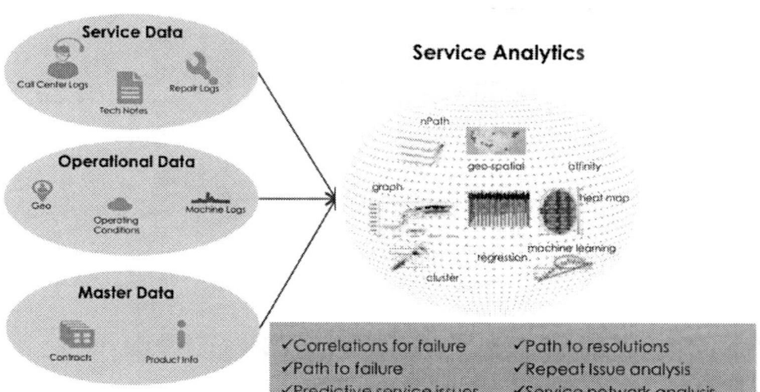

✓Correlations for failure
✓Path to failure
✓Predictive service issues
✓Early warning

✓Path to resolutions
✓Repeat Issue analysis
✓Service network analysis
✓Service profitability analysis

BI/Reporting
✓SLA Adherence
✓Comparative Analysis
✓Customer Patterns

Service Improvements
✓Resolution Guide
✓Product Quality
✓Proactive Fixes

Service Offers
✓Service offer redesign
✓Offer upsell
✓Offer cross-sell

Service Planning
✓Tech Resource Planning
✓Spares Planning
✓Service Network Planning
✓What-if Analysis
✓Capable to Service Analysis

Service 360: Business Benefits

- Proactive issue identification
- Faster & accurate issue resolutions

- Lower cost to serve
- Higher service profitability
- Increase in revenue through cross-sell and up-sell

- Higher net promoter score
- Customer satisfaction & retention

Sample use cases

- Organizations supporting data center temperature control and uptime
- Organizations leasing heavy equipment for production purposes (O&G, Mining)
- Organizations selling enterprise servers with tight support contracts

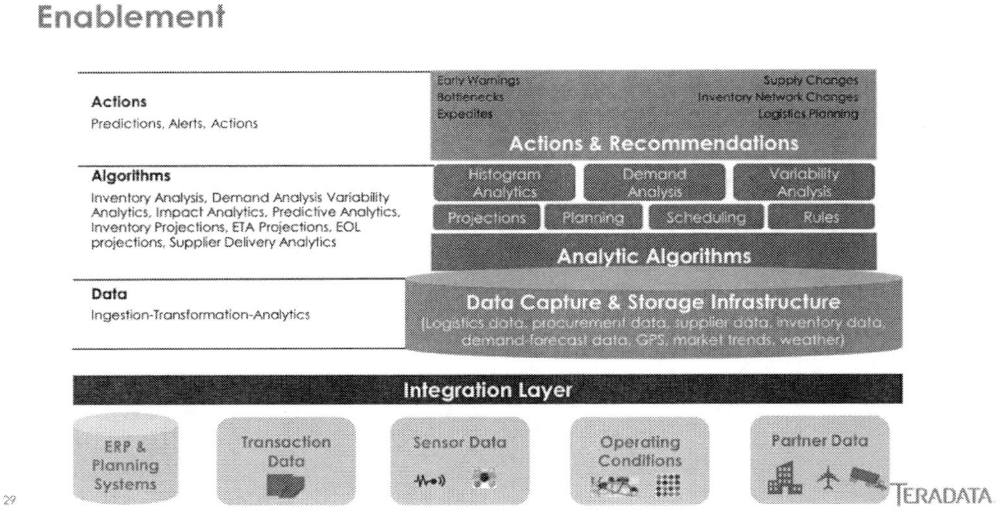
Enablement
Actions
Predictions, Alerts, Actions
Algorithms
Inventory Analysis, Demand Analysis Variability Analytics, Impact Analytics, Predictive Analytics, Inventory Projections, ETA Projections, EOL projections, Supplier Delivery Analytics
Data
Ingestion-Transformation-Analytics
Early Warnings
Bottlenecks
Expedites
Supply Changes
Inventory Network Changes
Logistics Planning
Actions & Recommendations
Histogram Analytics
Demand Analysis
Variability Analysis
Projections
Planning
Scheduling
Rules
Analytic Algorithms
Data Capture & Storage Infrastructure
(Logistics data, procurement data, supplier data, inventory data, demand-forecast data, GPS, market trends, weather)
Integration Layer
ERP & Planning Systems
Transaction Data
Sensor Data
Operating Conditions
Partner Data
29
TERADATA

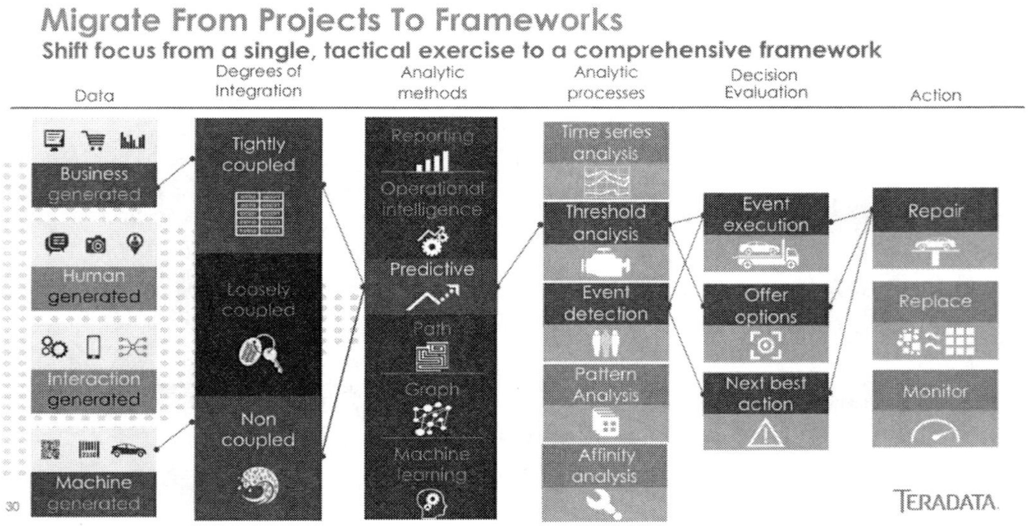
Migrate From Projects To Frameworks
Shift focus from a single, tactical exercise to a comprehensive framework
Data
Degrees of Integration
Analytic methods
Analytic processes
Decision Evaluation
Action
Business generated
Human generated
Interaction generated
Machine generated
Tightly coupled
Loosely coupled
Non coupled
Reporting
Operational intelligence
Predictive
Path
Graph
Machine learning
Time series analysis
Threshold analysis
Event detection
Pattern Analysis
Affinity analysis
Event execution
Offer options
Next best action
Repair
Replace
Monitor
30
TERADATA

TD Differentiators

Outcome based

- ✧ Actionable recommendations
- ✧ Monitors key metrics
- ✧ Incorporates measure, monitor & feedback loops

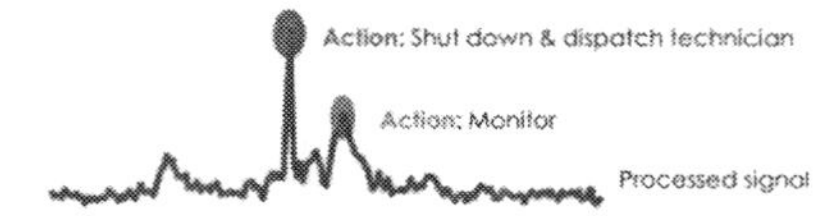

Standardized

- ✧ Specific business problems
- ✧ Proven Approach
- ✧ Data quality management
- ✧ Data models/dictionaries
- ✧ APIs/Analytical Models
- ✧ User Interface

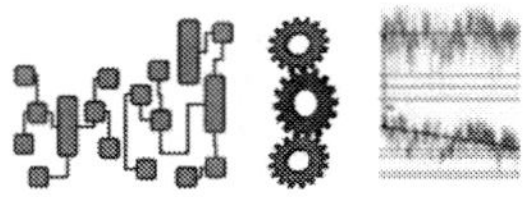

Configurable, Robust & Scalable

- ✧ Business rules can be configured
- ✧ Scalable environment
- ✧ Establishes a flexible environment to solve to other business problems
- ✧ Adapts to multiple technologies
- ✧ Highly Secure

TERADATA

Concluding Remarks

- You've learnt some new buzz words: Industry 4.0, IoT, Big Data, analytics...and hopefully you understand what they mean and how these ideas impact you
- You've learnt that there is a data driven approach that can build on and enhance your current processes
- You have a sneaking suspicion that it might be possible to unlock more value from your data
- Some of this unlocking could be straightforward integration of data sets but some of it will require some thinking to understand what the data might be telling you (Discovery)
- You have some good ideas on what business value and impact you can drive with this data

TERADATA

EMI AT THE DOW CHEMICAL COMPANY

Mary Beth Seasholtz, Brad David
With acknowledgements to May Roca, Megan Miner

AIChE Conference
March 2017

DOW 1

Author

Brad David

- 30 years of Dow experience (Manufacturing, Process Research, TS&D Support)
- Technical Leader for Manufacturing Analytics
- Data Analytics for ~8 years

DOW 2

Outline

- Why the EMI Journey at Dow
- What has been done so far
- How we have implemented and supported EMI
  - Technically
  - Culturally
- Future of EMI at Dow

Dow

Dow and EMI journey

The Numbers

- Since 2012, we have successful deployed over 60 EMI implementations in a variety of manufacturing environments
- There is ~500 Dow users

Why?

- Take advantage of the vast amount of data currently being stored and unused - "Big Data"
  - Organize and manage Dow's data more effectively
- There is value to be gained from data
  - Drive down Operational Costs
  - Deliver on Productivity Improvements (i.e. AU)

Dow

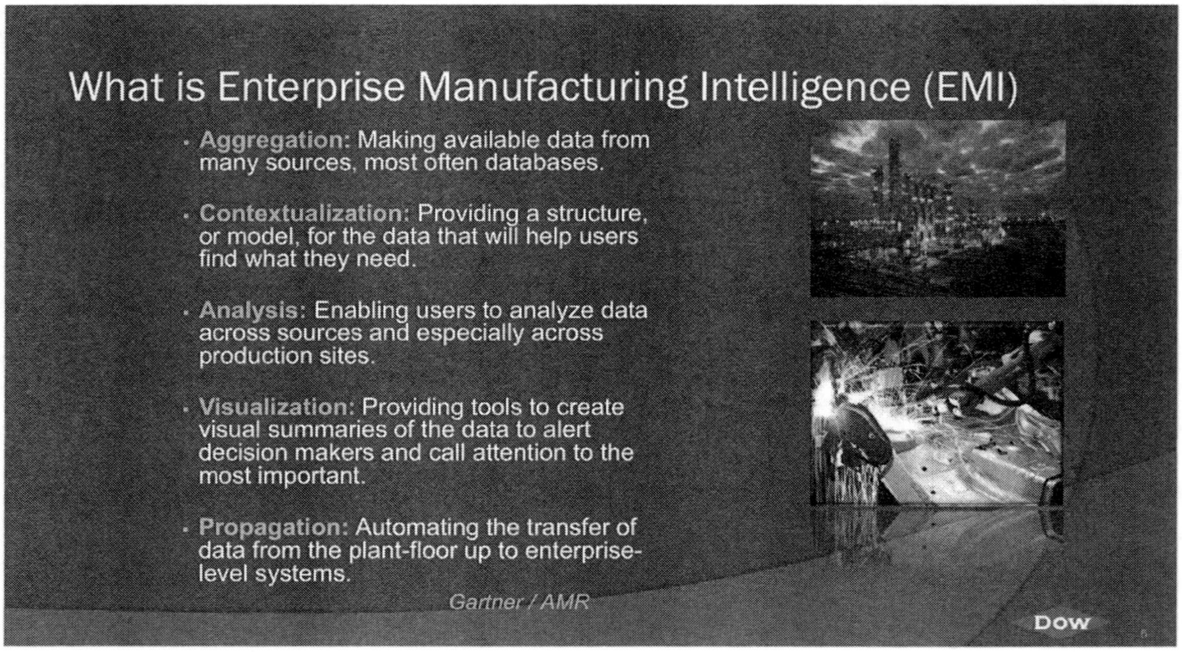
What is Enterprise Manufacturing Intelligence (EMI)
Aggregation: Making available data from many sources, most often databases.
Contextualization: Providing a structure, or model, for the data that will help users find what they need.
Analysis: Enabling users to analyze data across sources and especially across production sites.
Visualization: Providing tools to create visual summaries of the data to alert decision makers and call attention to the most important.
Propagation: Automating the transfer of data from the plant-floor up to enterprise-level systems.
Gartner / AMR
Dow

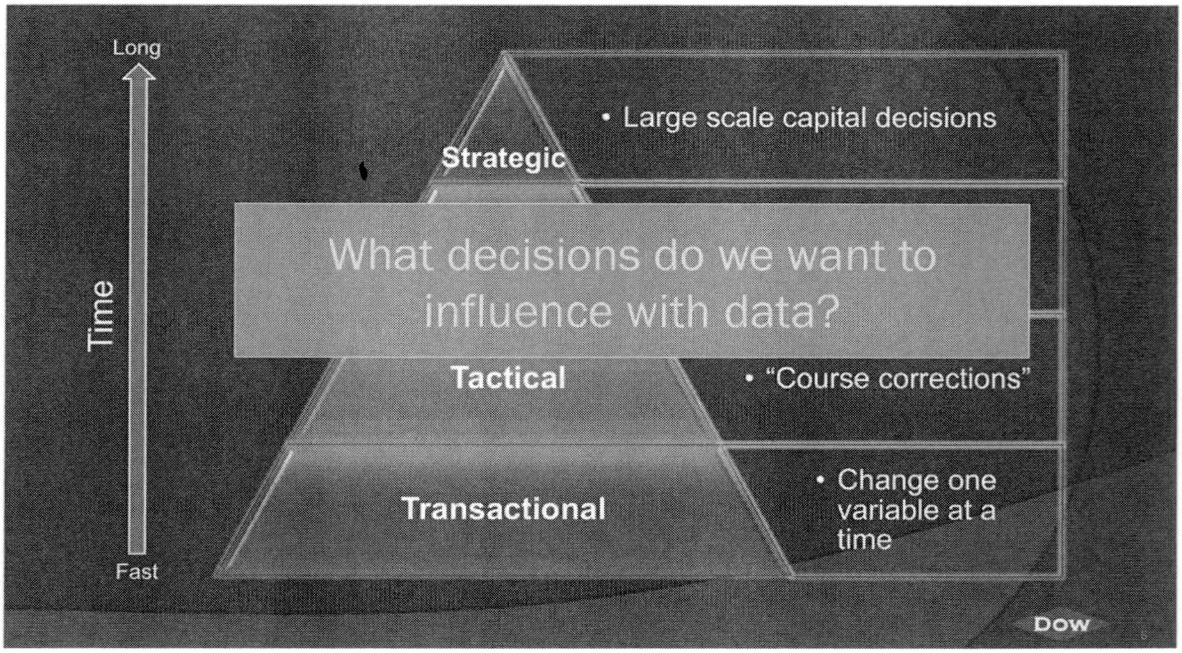
Long
Time
Fast
Strategic
Large scale capital decisions
What decisions do we want to influence with data?
Tactical
"Course corrections"
Transactional
Change one variable at a time
Dow

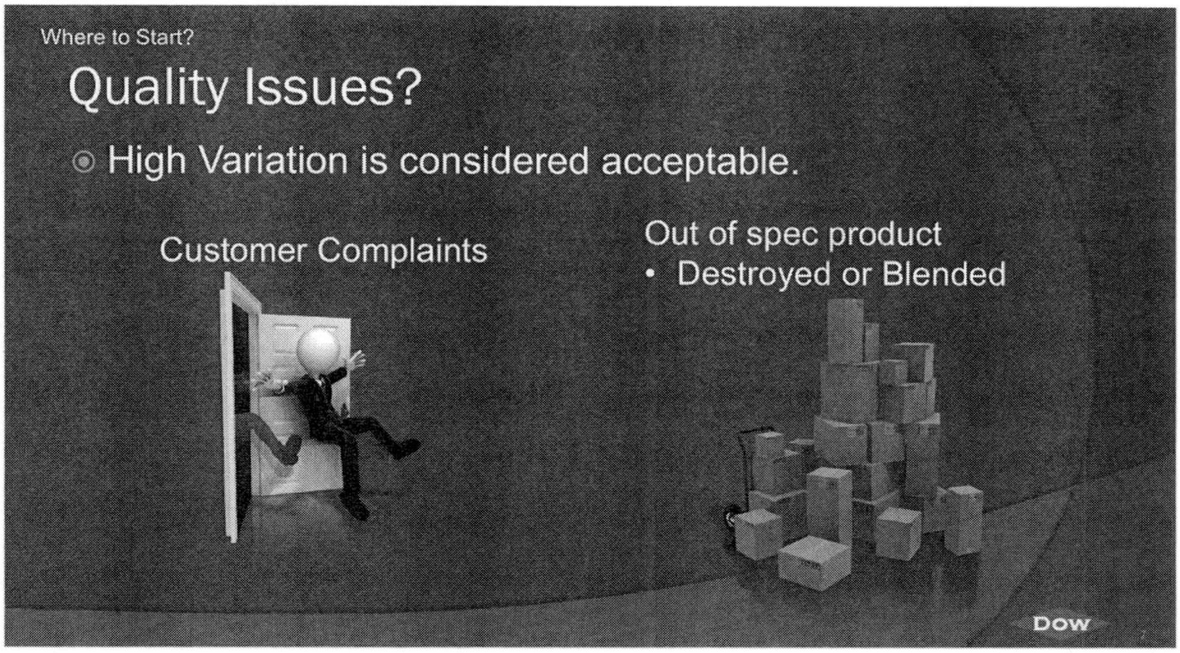
Where to Start?
Quality Issues?
High Variation is considered acceptable.
Customer Complaints
Out of spec product
• Destroyed or Blended
DOW

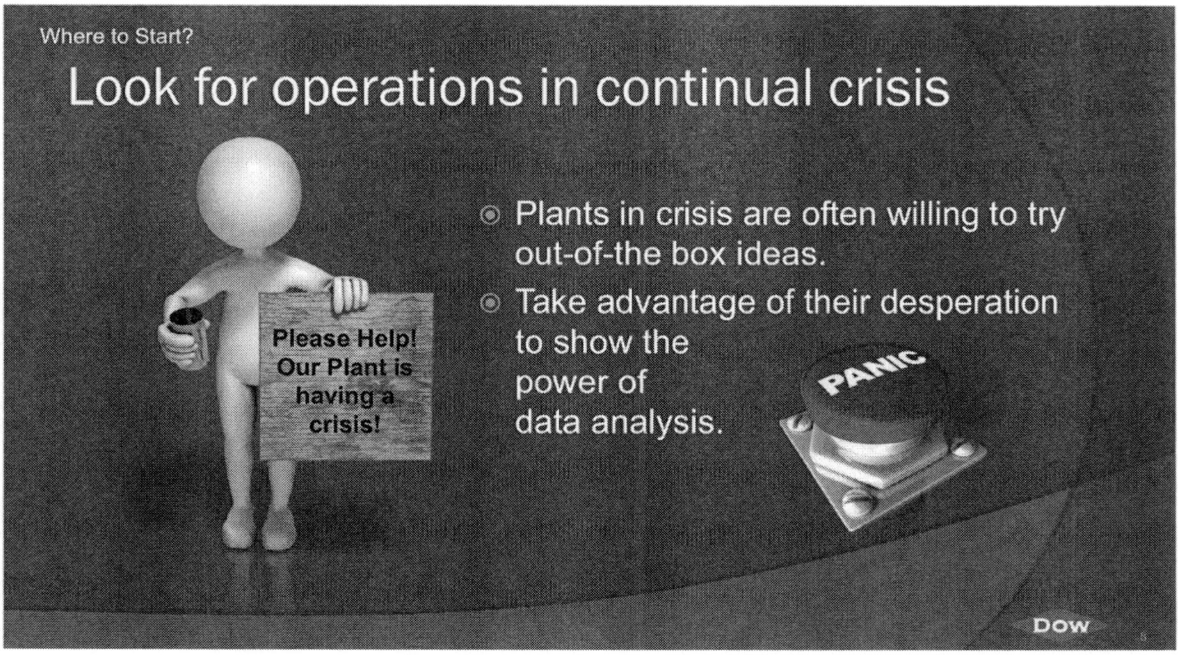
Where to Start?
Look for operations in continual crisis
Plants in crisis are often willing to try out-of-the box ideas.
Take advantage of their desperation to show the power of data analysis.
Please Help! Our Plant is having a crisis!
PANIC
DOW

Where to Start?
When the storm has passed
Whispers of Local Lore: "Let me tell you how we used to do it. I know it is better than what we're doing now."
Problem Solved!
Let's never do THAT again!
Learnings
Use DATA to refocus!
- Possibly as a Control Plan
DOW
9

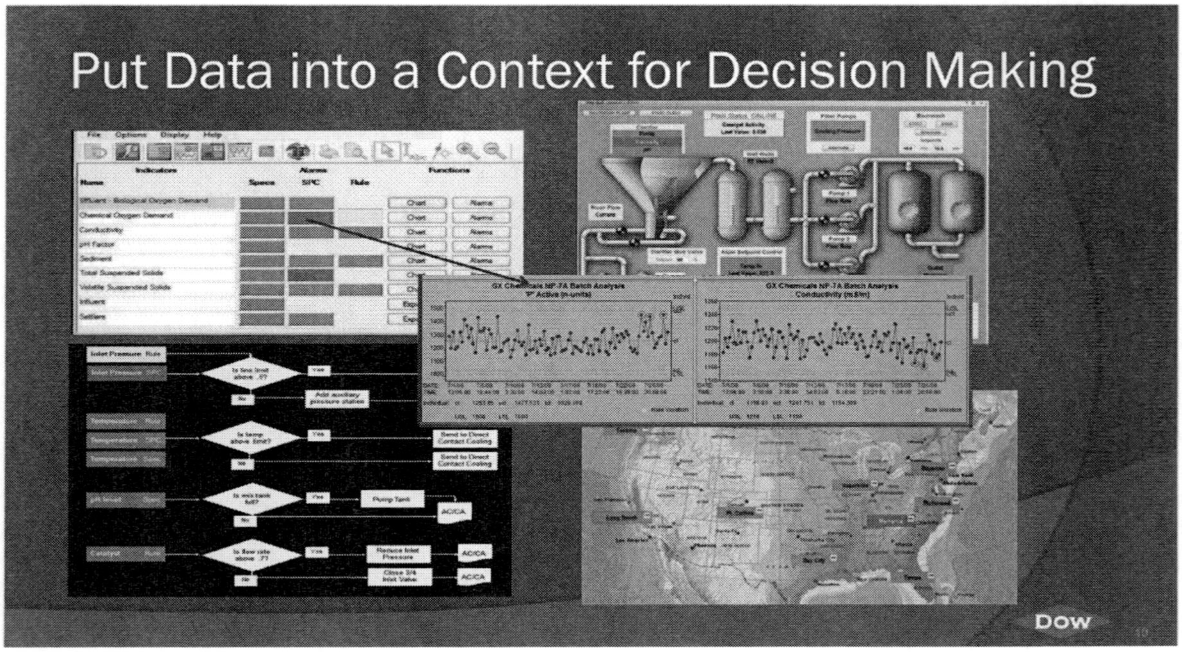
Put Data into a Context for Decision Making
DOW
10

Put Data and Decision Making Together

- Our role is to guide the conversation and translate into data/action.
- Get technical experts to tell us what they don't realize they know!

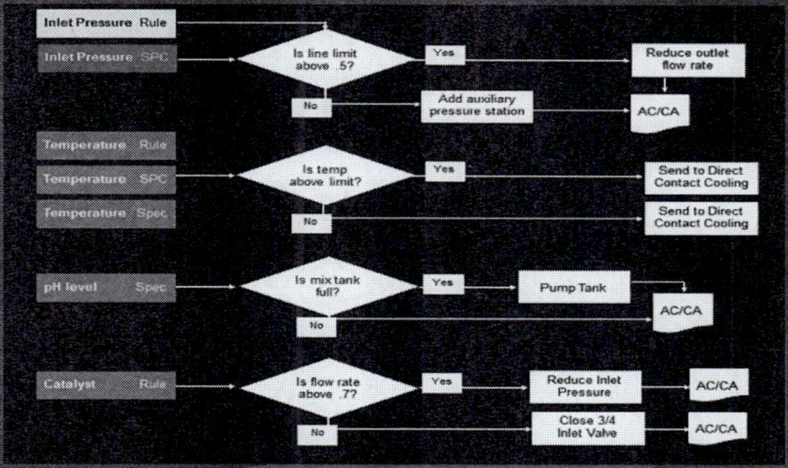

DOW

Successes

Manage Laboratory Data

Opportunity
- Hard to keep an eye on 264 Control Charts across 55 instruments and 2 labs

EMI
- Automated monitoring of Lab Verification Data.
- Solidified Operating Discipline

Key Results
- Proactive vs. Reactive
- 7 vs. 24 hrs/ month to maintain instrument verification.

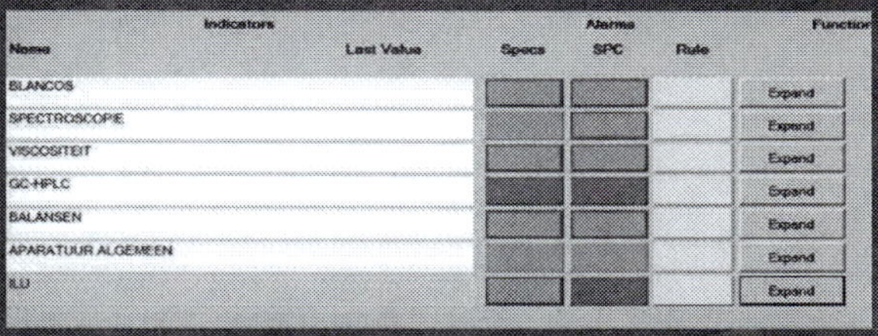

DOW

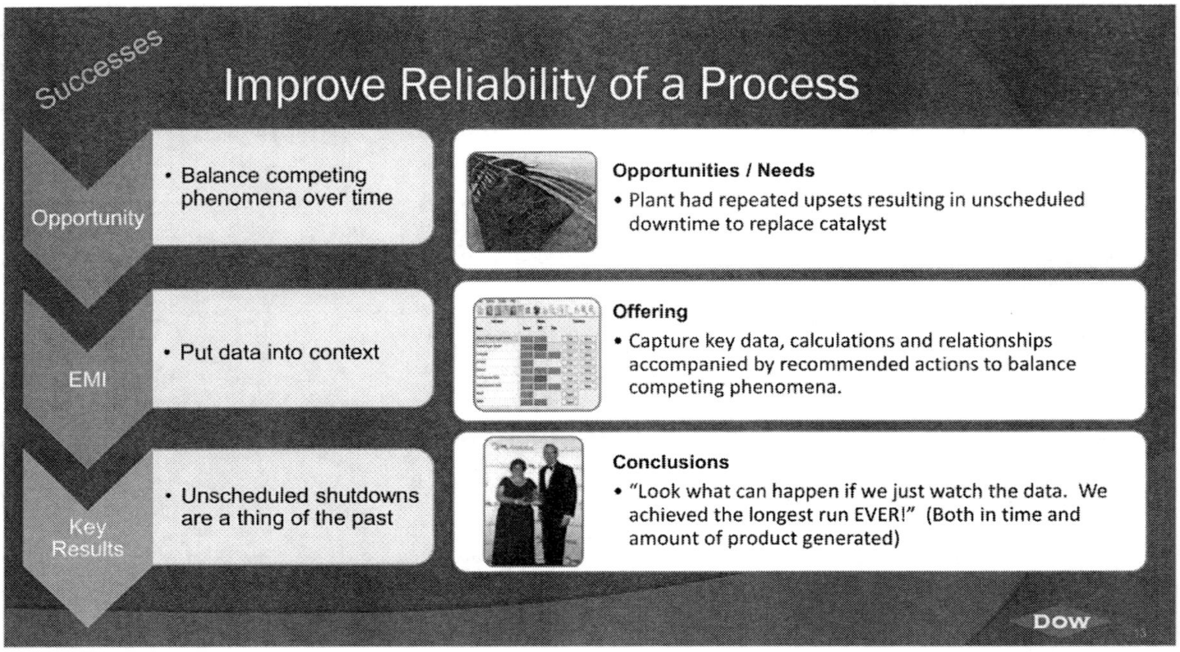
Successes
Improve Reliability of a Process
Opportunity
• Balance competing phenomena over time
EMI
• Put data into context
Key Results
• Unscheduled shutdowns are a thing of the past
Opportunities / Needs
• Plant had repeated upsets resulting in unscheduled downtime to replace catalyst
Offering
• Capture key data, calculations and relationships accompanied by recommended actions to balance competing phenomena.
Conclusions
• "Look what can happen if we just watch the data. We achieved the longest run EVER!" (Both in time and amount of product generated)
Dow

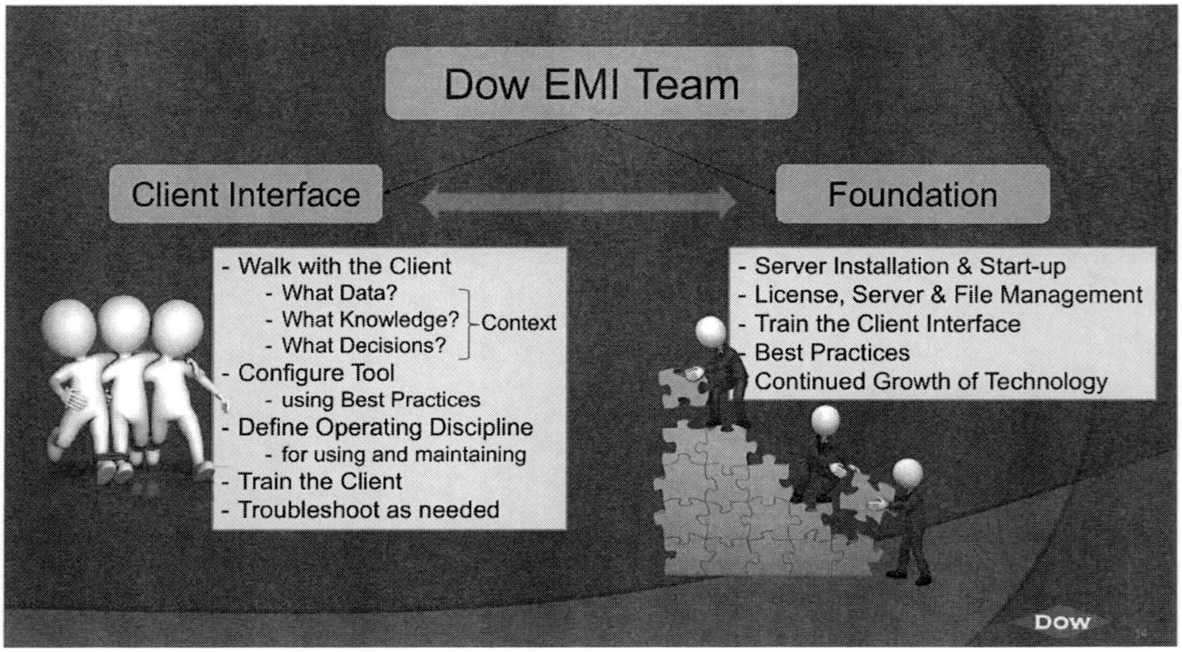
Dow EMI Team
Client Interface
Foundation
- Walk with the Client
- What Data?
- What Knowledge?
- What Decisions?
Context
- Configure Tool
- using Best Practices
- Define Operating Discipline
- for using and maintaining
- Train the Client
- Troubleshoot as needed
- Server Installation & Start-up
- License, Server & File Management
- Train the Client Interface
- Best Practices
- Continued Growth of Technology
Dow

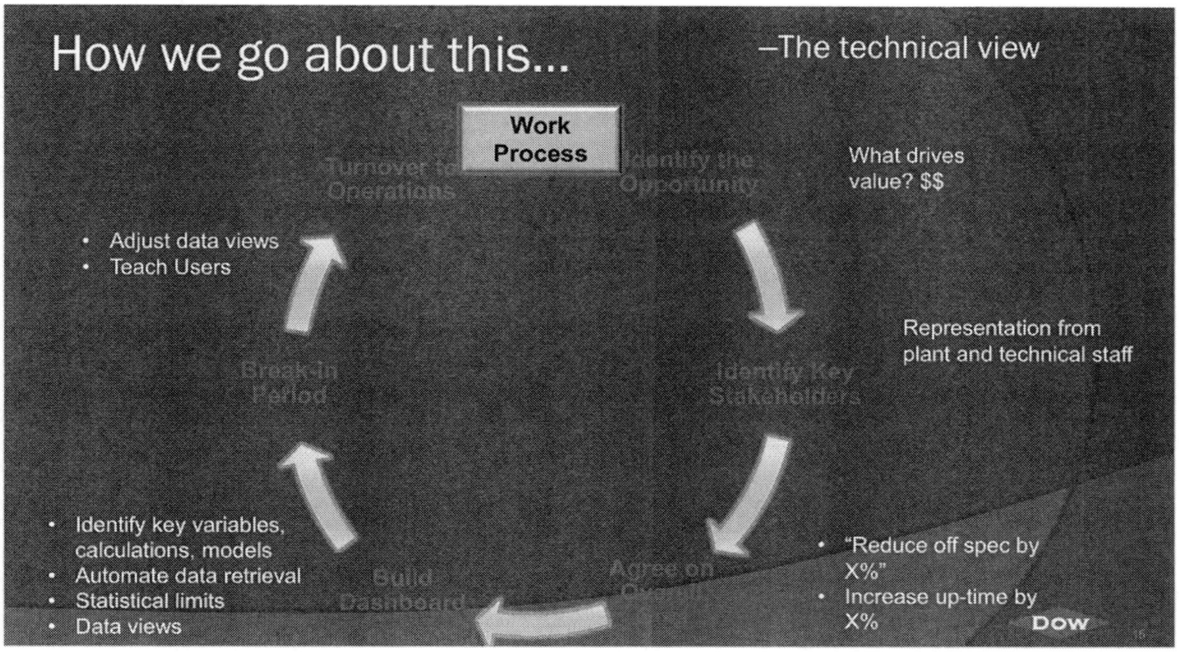
How we go about this...
–The technical view
Work Process
Turnover to Operations
Identify the Opportunity
What drives value? $$
Adjust data views
Teach Users
Representation from plant and technical staff
Break-in Period
Identify Key Stakeholders
Identify key variables, calculations, models
Automate data retrieval
Statistical limits
Data views
Build Dashboard
"Reduce off spec by X%"
Increase up-time by X%
DOW

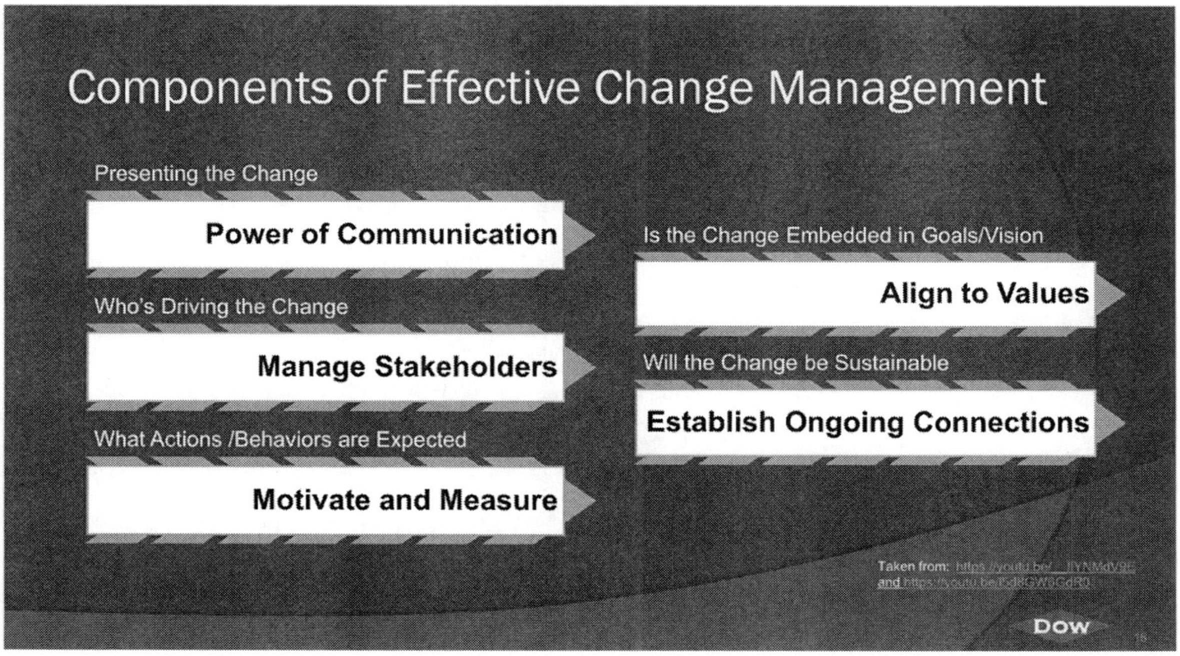
Components of Effective Change Management
Presenting the Change
Power of Communication
Is the Change Embedded in Goals/Vision
Align to Values
Who's Driving the Change
Manage Stakeholders
Will the Change be Sustainable
Establish Ongoing Connections
What Actions /Behaviors are Expected
Motivate and Measure
Taken from:
DOW

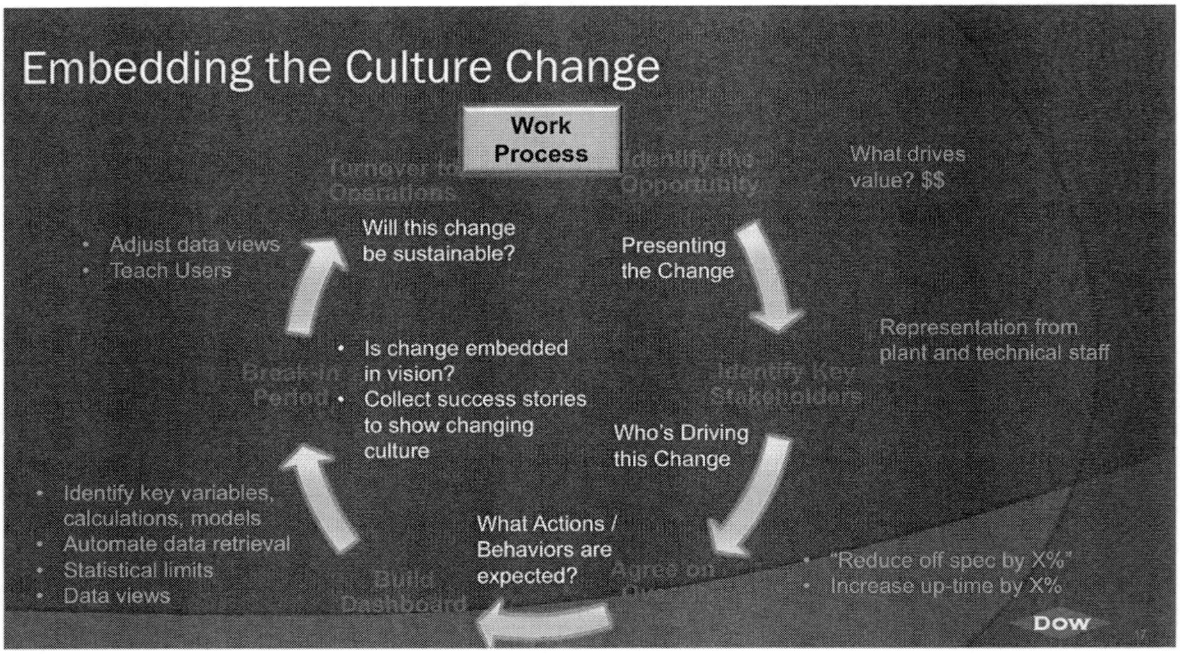
Embedding the Culture Change
Work Process
Identify the Opportunity
What drives value? $$
Presenting the Change
Identify Key Stakeholders
Representation from plant and technical staff
Who's Driving this Change
"Reduce off spec by X%"
Increase up-time by X%
What Actions / Behaviors are expected?
Build Dashboard
Identify key variables, calculations, models
Automate data retrieval
Statistical limits
Data views
Break-in Period
Is change embedded in vision?
Collect success stories to show changing culture
Turnover to Operations
Will this change be sustainable?
Adjust data views
Teach Users
DOW

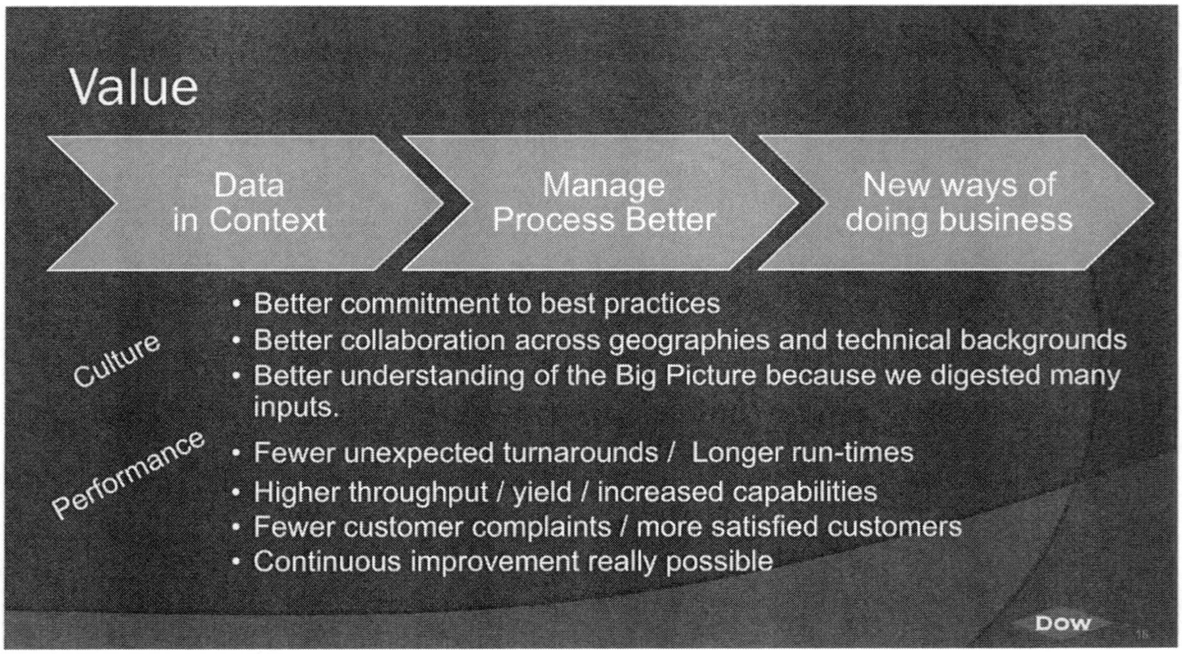
Value
Data in Context
Manage Process Better
New ways of doing business
Culture
Better commitment to best practices
Better collaboration across geographies and technical backgrounds
Better understanding of the Big Picture because we digested many inputs.
Performance
Fewer unexpected turnarounds / Longer run-times
Higher throughput / yield / increased capabilities
Fewer customer complaints / more satisfied customers
Continuous improvement really possible
DOW

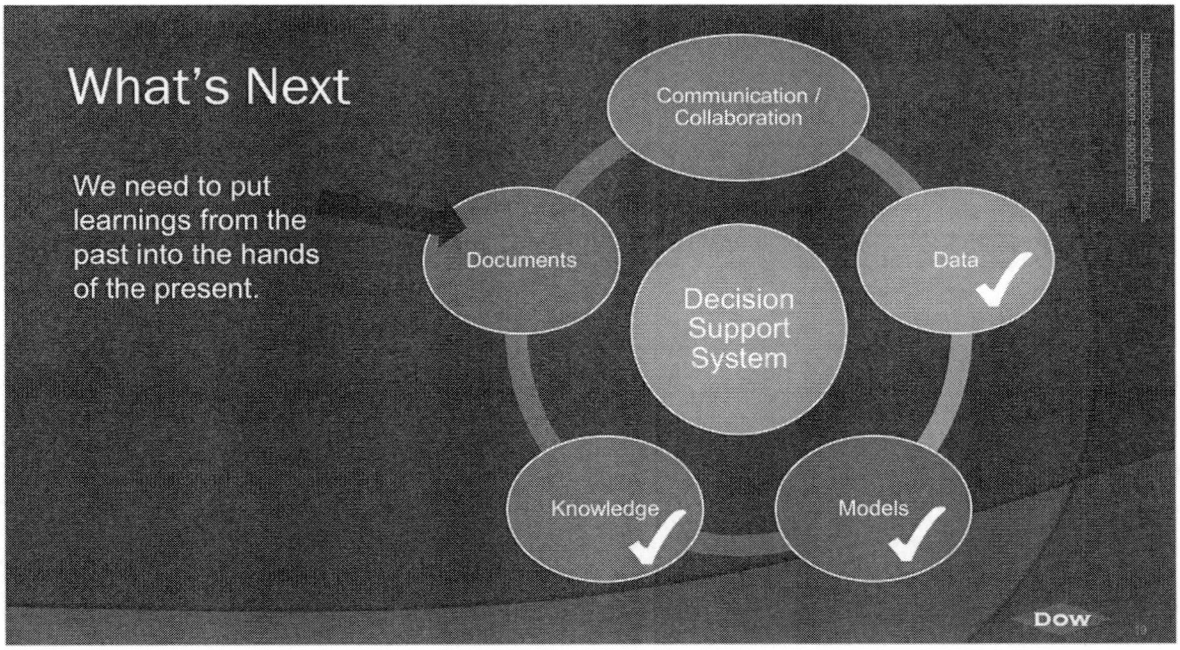

Conclusions

- We have completed 60 deployments in a variety of manufacturing environments since 2012.
  - *"There is a lot of money on the table."*
- We have a group that
  - Helps clients put data into context for decision making
    - A technical and cultural journey
  - Care and Feeding of an IT Foundation

DOW 20

EMI AT THE DOW CHEMICAL COMPANY

Mary Beth Seasholtz, Brad David

With acknowledgements to May Roca, Megan Miner

AIChE Conference
Mar 2017

THANK-YOU!

DOW

21

479791 Visualization and Visual Analysis of Batch-Continuous Process Data

Abstract:

The operation of batch-continuous processes is a unique case as it contains the characteristics of both batch and continuous processes. The retention of characteristics unique to each type of process while also including interactions between the sub-processes render the operation of batch-continuous process an interesting topic of study.

While the design[1,2] and scheduling[3,4] of batch-continuous processes have received some attention in the literature, the monitoring of such processes remains unstudied. We will discuss a novel geometric method for data-driven monitoring of batch-continuous processes in this presentation. Our work is an extension of the time-explicit Kiviat (radial) plot visualization and fault detection frameworks that were previously developed for both continuous processes[5] as well as batch processes[6]. In this framework, each sample data point of the multivariate time-series collected from process operations is represented in a radial plot. These plots are stacked vertically in the order they were acquired, resulting in a time-explicit representation of the multivariate time series. The geometric properties of this setup allow for the process state at a given time to be represented as a single point, the centroid of the corresponding polygon in radial coordinates.

To extend these ideas to batch-continuous processes, we propose a parallel monitoring fault detection algorithm, whereby fault detection for the whole process is done at any given time by executing both continuous and batch fault detection algorithms in parallel. Based on the fault detection results of both algorithms we are able to identify if a fault exists in the process, and if so, which part of the process (batch or continuous) is the fault located in.

We will also introduce the use of classification techniques that will be used to split process data into batch and continuous process training data for the generation of the appropriate confidence regions. We demonstrate our methodology on a simulation case study of a batch-continuous process.

References

1. Knopf, F. Carl, Martin R. Okos, and Gintaras V. Reklaitis. "Optimal design of batch/semicontinuous processes." ***Industrial & Engineering Chemistry Process Design and Development*** 21.1 (1982): 79-86.
2. Yeh, N. C., and G. V. Reklaitis. "Synthesis and sizing of batch/semicontinuous processes: single product plants." ***Computers & chemical engineering*** 11.6 (1987): 639-654.
3. Ierapetritou, M. G., and C. A. Floudas. "Effective continuous-time formulation for short-term scheduling. 2. Continuous and semicontinuous processes." ***Industrial & engineering chemistry research*** 37.11 (1998): 4360-4374.
4. Karimi, Iftekhar A., and Conor M. McDonald. "Planning and scheduling of parallel semicontinuous processes. 2. Short-term scheduling." ***Industrial & Engineering Chemistry Research*** 36.7 (1997): 2701-2714.
5. Wang, R. C., Edgar, T. F., Baldea, M., Nixon, M., Wojsznis, W., & Dunia, R. (2015). Process fault detection using time explicit Kiviat diagrams. ***AIChE Journal, 61***(12), 4277-4293
6. Wang, R., Baldea, M., & Edgar, T.F. (2015). Visualization and Data-Driven Monitoring of Batch Processes. ***2015 AIChE Annual Meeting,*** 418567

480399 Regression Strategies for Large Data Sets

Abstract:

ABSTRACT for AIChE 2017 Spring Meeting

Regression Strategies for Large Data Sets

James C Cross III

The MathWorks, Inc.

james.cross@mathworks.com

617-605-5818

Plant operators and engineers are increasingly using historian data to gain insight into the relationships between and amongst process parameters and product attributes. A universal goal is the determination of the operating conditions that produce the highest product output and/or quality per unit cost. Models provide insight, but seldom account for myriad physical nuances which sometimes exhibit significant influence.

The quantity of data that must be analyzed in pursuit of the sought inferences is invariably large, consisting of hundreds or thousands of quantities sampled at many millions of points in time. It is seldom possible to process data sets of this size in computing resource memory.

A number of frameworks for partitioning large data sets have been developed and popularized, though adoption by industrial companies has been notably slower than by IT-centric enterprises. This paper endeavors to demystify the processing of out-of-memory data for an engineering audience.

Regression, which is fundamental in data analysis, is used to motivate use of the techniques. This paper begins with computing simple statistics. Subsequently, some strategies for matrix manipulations (relevant to both direct and iterative methods) are discussed. Finally, an illustrative example of a large data set regression is presented.

480093 The Data Analytics Triangle

Abstract:

When analyzing real-world data sets, there is no one-size-fits-all algorithm. This talk discusses the importance of and methods for interrogating a dataset to understand its properties and better select appropriate data analytics methods. The presentation uses the framework of the "Data Analytics Triangle," which organizes datasets by associating each of the vertices of the triangle with one of three characteristics: nonlinearity, correlation, and dynamics. The edges of the triangle represent datasets that have two of these characteristics, whereas the center represents datasets that have all three characteristics. We then associate each of these parts of the triangle with different data analytics methods, so that the data integration and data analytics triangle guide the user to the best data analytics methods to apply for a particular data set. The final algorithm to apply then depends both and the properties of the dataset and the goal of the data-based model (e.g., prediction, classification).

479736 Advances in Batch Data Analysis

Abstract:

Batch processes are ubiquitous in modern industry, namely in the production of high valued products and fine chemicals, due to their added flexibility and scalability when compared to continuous processing systems. Therefore, they are often considered as priority targets for data-driven process improvement activities. These activities may aim at different goals, such as improving process safety, efficiency, environmental impact, product quality and economical/financial bottom line results of the company. According to the nature of the activities, the tasks may need to be carried out online or offline, and always in a robust and manageable way - manageable in the sense that they can be handled within the knowledge and skills resources likely to be available in companies (possibly after a period of technology transfer). This may be particularly challenging to achieve for batch processes, due to the confluence of several well-known elements of complexity, such as: the tensorial nature of data collected from these processes, having a with a 3-way structure, $\boldsymbol{X}$ ($\boldsymbol{I}$ x $\boldsymbol{J}$ x $\boldsymbol{K}$), where $\boldsymbol{I}$ is the number of batches, $\boldsymbol{J}$ is the number of variables and $\boldsymbol{K}$ is the number of measurements collected during a batch; their intrinsic dynamic and non-stationary nature; the occurrence of multiple stages, not all of them having equal importance in the final outcome; the existence of many variables involved, exhibiting a variety of typical time-profiles; among others.

In this tutorial, an overview will be provided about methodologies available to address data-driven process improvement of batch operations. Both online and offline activities will be targeted. A systematic way of organizing the different proposals available for conducting the analysis will also be provided, which may be instrumental for devising a roadmap for addressing the complexity of batch processes in more efficient and consistent way. The ultimate goal is fostering the exploitation of the information available from this type of processes in companies, by handling their complexity through the use of proper methodologies, within a clearly defined workflow.

480967 Performance Monitoring of Global Chemometric Models in Manufacturing Plants: New Approaches for Efficient Model Maintenance

Abstract:

Online FTIR is commonly employed for physical property and composition analysis in manufacturing plants. In general, the most common approach involves utilization of local, site-specific chemometric models to predict properties that would otherwise require labor intensive offline analysis. This approach can only be efficiently used for a few products. Performance monitoring in this type of setting is routinely carried out by periodically collecting audit samples and comparing the model prediction to a primary analysis. The choice of products to be monitored is a straight forward process due to the limited scope. Results of such auditing processes are key for making maintenance decisions from a hardware or chemometric model perspective.

Dow's Polyethylene business aims for a consistent global quality assurance approach with globally harmonized analytical tools and capabilities. Part of this strategy approach requires the creation of chemometric models that can be successfully employed for continuous analysis of multiple properties and products on all production trains globally. In such a scenario, traditional model maintenance approaches fail to produce an accurate maintenance plan that can facilitate proactive and targeted use of resources for improved future model performance. In order to overcome these shortcomings, various tools focused on quality, equivalency and capability have been developed. This presentation will discuss the designed audit program system that allows alignment of information collected using the traditional approach with process data that can provide additional insight regarding model prediction performance. Detailed analysis enables early identification of the biggest opportunities for improvement and maintenance.

2017 AIChE®

SPRING MEETING &

13TH

GLOBAL CONGRESS

ON PROCESS SAFETY

March 26–30, 2017
Henry B. Gonzalez
Convention Center
San Antonio, TX

17 AIChE Spring Meeting, San Antonio, TX

13TH GLOBAL CONGRESS ON PROCESS SAFETY

---

17 AIChE Spring Meeting, San Antonio, TX

13TH GLOBAL CONGRESS ON PROCESS SAFETY

IIoT for Process Control

- George Buckbee
  - Head of Performance Solutions at Metso
  - Fellow of the International Society of Automation

2017 AICHE® SPRING MEETING & 13TH GLOBAL CONGRESS ON PROCESS SAFETY

Overview

- What is IIoT?
- Application of IIoT to Process Control Systems
- Examples: What can be done today?
- What does the future hold?

What is IIoT?

- IIoT = Industrial Internet of Things
  = "The application of IoT to Manufacturing"
- IoT = "The internetworking of smart devices to exchange data"
- Microsoft's High-Level Description of IoT Services:

Devices | Connections | History | Analytics | Presentation

17AIChE Spring Meeting, San Antonio, TX — 13TH GLOBAL CONGRESS ON PROCESS SAFETY

Do we have IoT in Process Control?

Devices	Connections	History	Analytics	Presentation
Control Loops & Processes	Many other connections	Process Data Historians	Offline Analysis	Solution-Dependent Tools
Augmented by Context	ODBC/SQL	Laboratory Info Mgmt Systems	Home-Grown Tools	
Not-so-smart Instruments, Valves, etc.	xml	Computerized Maintenance Management	Some Vendor Solutions	
Smart Instruments, Valves, etc.	OPC			Back to DCS or HMI
	HART			
	Fieldbus			

17AIChE Spring Meeting, San Antonio, TX — 13TH GLOBAL CONGRESS ON PROCESS SAFETY

Does IIoT Require the Cloud ?

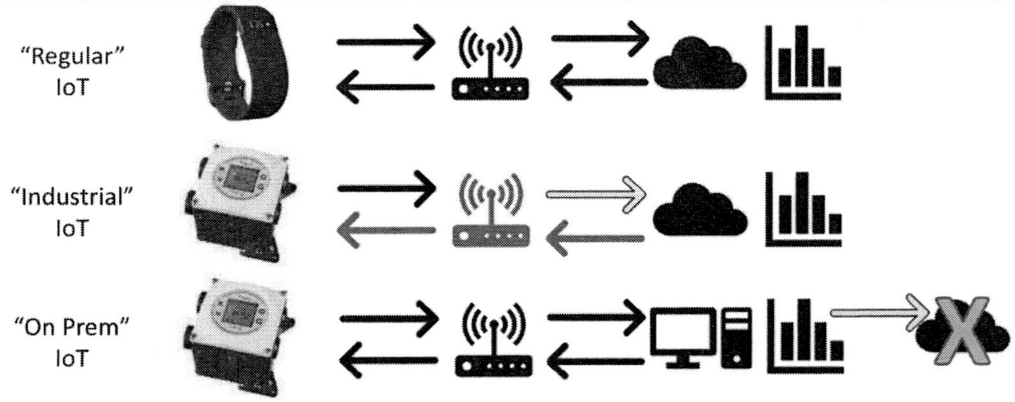

2017 AIChE® SPRING MEETING & 13TH GLOBAL CONGRESS ON PROCESS SAFETY

17AIChE Spring Meeting, San Antonio, TX 13TH GLOBAL CONGRESS ON PROCESS SAFETY

Where is there potential?

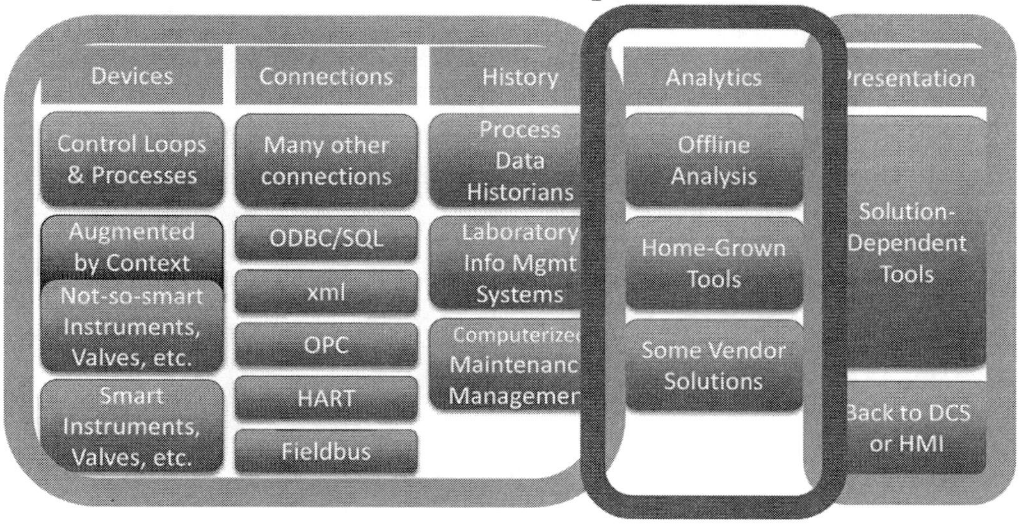

17AIChE Spring Meeting, San Antonio, TX 13TH GLOBAL CONGRESS ON PROCESS SAFETY

Examples: IIoT for Process Control

Instrument Health

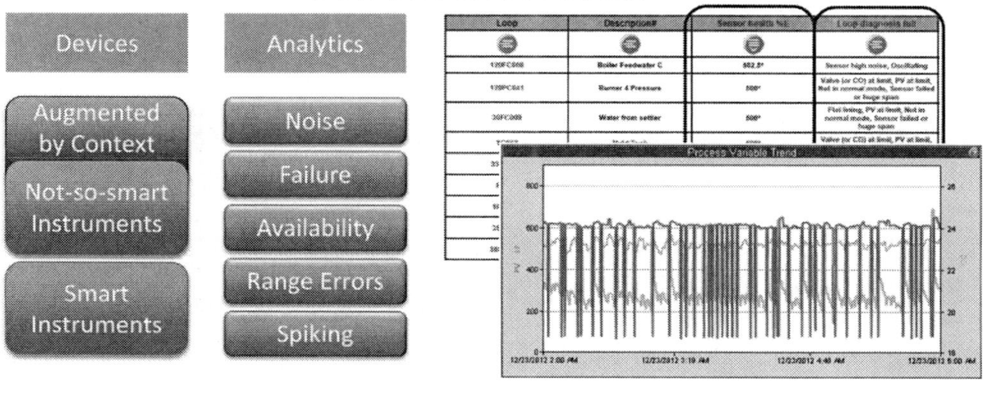

Examples: IIoT for Process Control

Valve Health

Devices	Analytics
Augmented by Context	Sizing
Not-so-smart Valves	Stiction
Smart Valves	Excess Travel
	Air Consumed
	Spiking

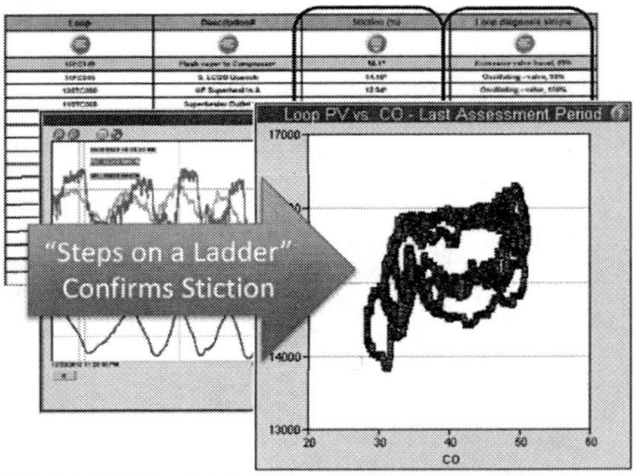

17 AIChE Spring Meeting, San Antonio, TX
13TH GLOBAL CONGRESS ON PROCESS SAFETY

Examples: IIoT for Process Control

Loop Tuning

Devices	Analytics
DCS & PLCs	Variance
Regulatory Loops	Need Tuning?
Cascades	Model
APC/MPC	Error
	Interaction

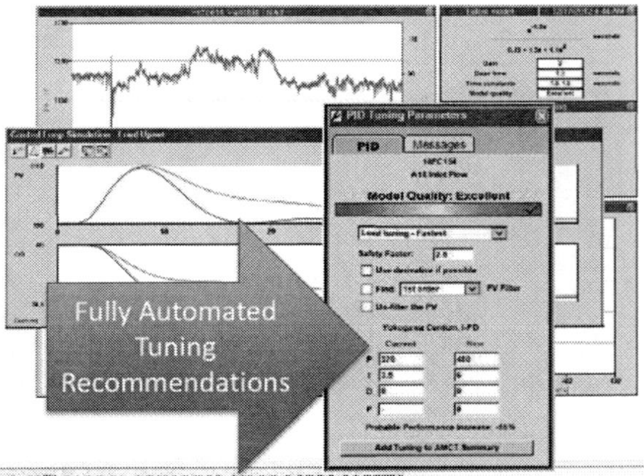

17AIChE Spring Meeting, San Antonio, TX — 13TH GLOBAL CONGRESS ON PROCESS SAFETY

Examples: IIoT for Process Control

Root Cause Analysis

Devices	Analytics
Processes	Interactions
Instruments	Strength
Controls	Time Shift
Analyzers	Root Cause
Soft Sensors	Model

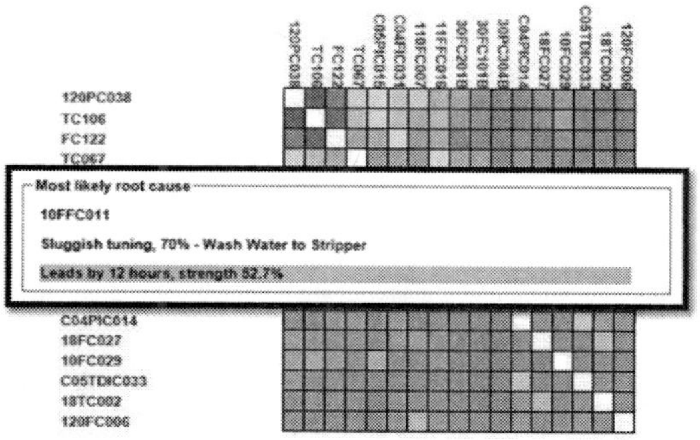

17AIChE Spring Meeting, San Antonio, TX — 13TH GLOBAL CONGRESS ON PROCESS SAFETY

What is the benefit of IIoT for Process Control?

- Leverages Process Data
- Leverages Engineering Time
- Improved Control
- Improved Maintenance Planning
- Improved Stability / Operation
- Reduced Cost of Operation
- Increased Production
- Improved Quality
- Results Depend Upon Where You Focus the Tools

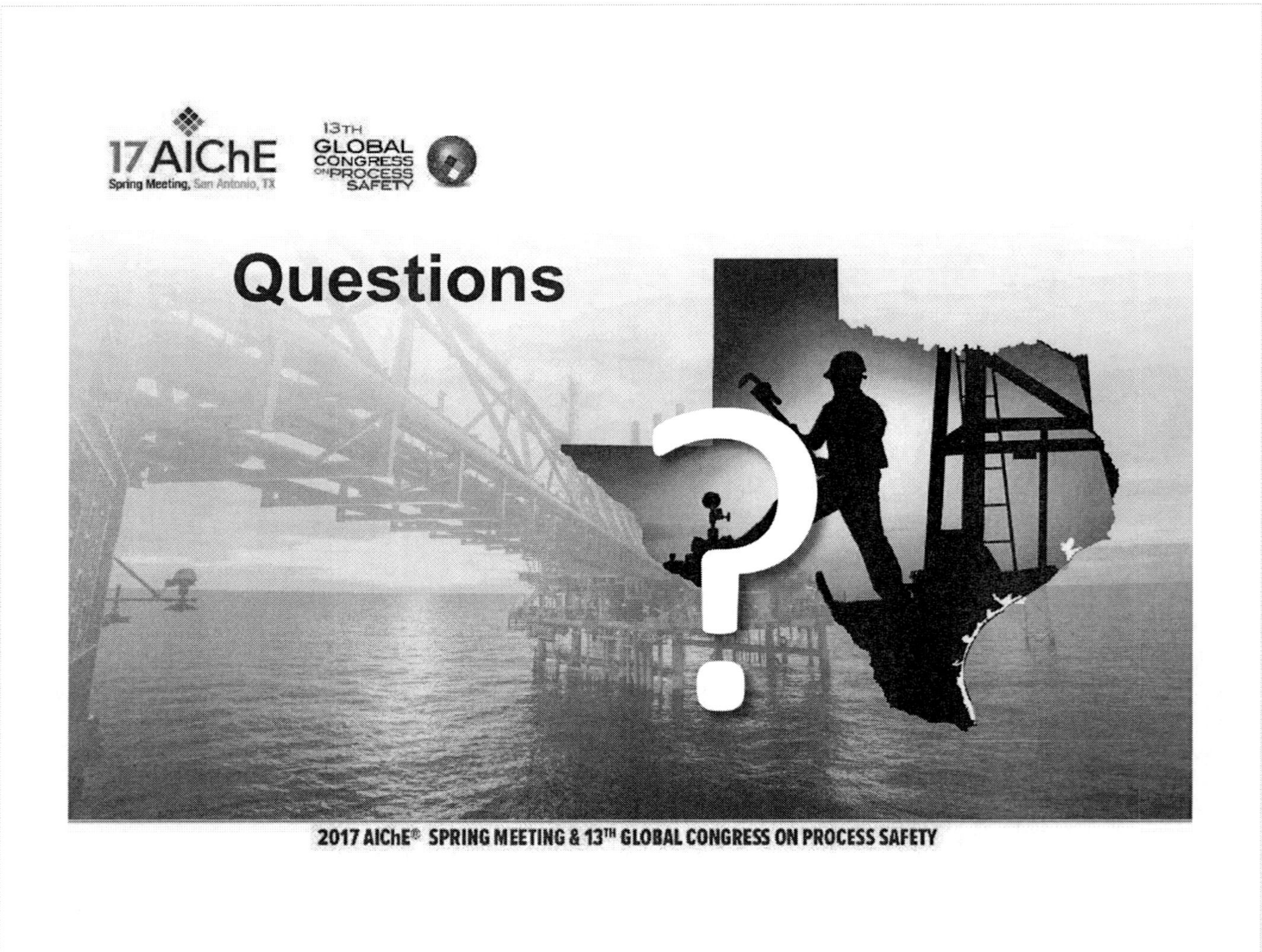
17 AIChE
Spring Meeting, San Antonio, TX
13TH GLOBAL CONGRESS ON PROCESS SAFETY
Questions
?
2017 AIChE® SPRING MEETING & 13TH GLOBAL CONGRESS ON PROCESS SAFETY

Knowledge discovery and explanation from process data using clustering and subspace search

Wenbo Zhu, Jose A. Romagnoli
Department of Chemical Engineering
Louisiana State University

LSU

1

Outline

- **Background**
  - Big data and process monitoring
- **Knowledge discovery**
  - Dimensional reduction
  - Data clustering
  - Cluster explanation
  - ***FASTMAN*** automatic data mining tool
- **Applications**
  - Industrial Pyrolysis Reactor
- **Concluding remarks**

2

Big data and Machine Learning

Machine learning gives computers the ability to learn without being explicitly programmed. Explores the study and construction of algorithms that can learn from and make predictions on data

- "***Data rich***" era
- Use data to gain "***knowledge***"
- Heavily used in social media: Facebook, Netflix, Google use Machine Learning and data mining

- Modern chemical plants store large amount of process data measurements

Use historical data for process monitoring and fault detection

3

Knowledge discovery

- The best diagnosis methods are *supervised*
  - Require data separated into "normal" and "fault" groups

However,

- Separated data is seldom available in practice
  - Advantage of data based monitoring: expedient modeling

Novel data clustering and dimensionality reduction tools drawn from computer science literature can assist in extracting knowledge from process data

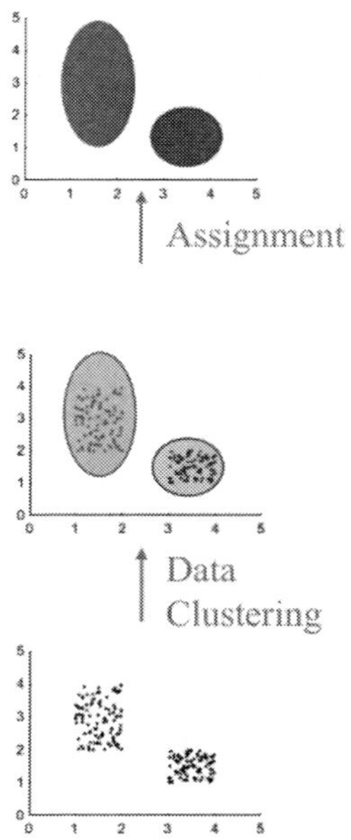

4

Dimensionality Reduction

- **Curse of dimensionality**: distances become less meaningful as number of dimensions grow
- **Reduced dimensional data**:
  - Easier to visualize
  - Removal of correlated information
- **Techniques Samples**:
  - No DR: cluster normalized data
  - PCA: standard matrix decomposition and projection
  - ICA: decomposition to stat. independent comps.
  - Kernel PCA: kernel trick for nonlinear projection
  - Isomap: uses graph structure for projection
  - Spectral Embedding (SE): preserved local topology of data
- **Essential for visualizing high dimensional data**

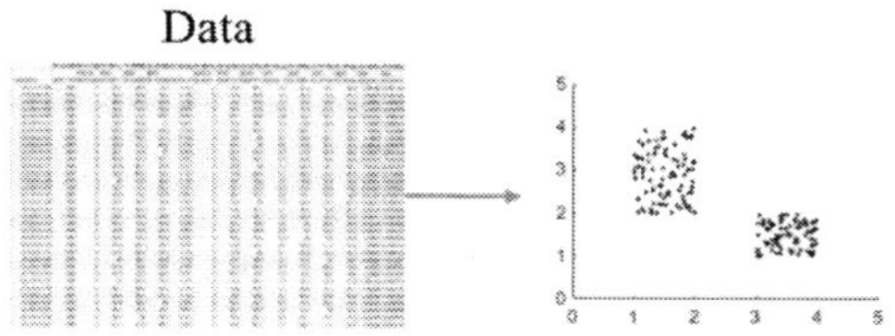

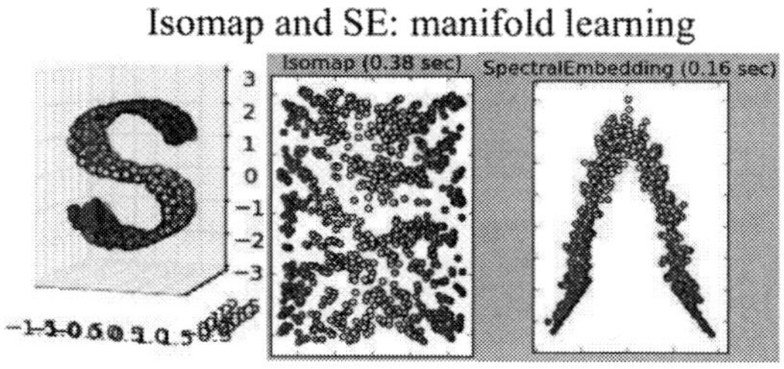

5

Dimensionality Reduction

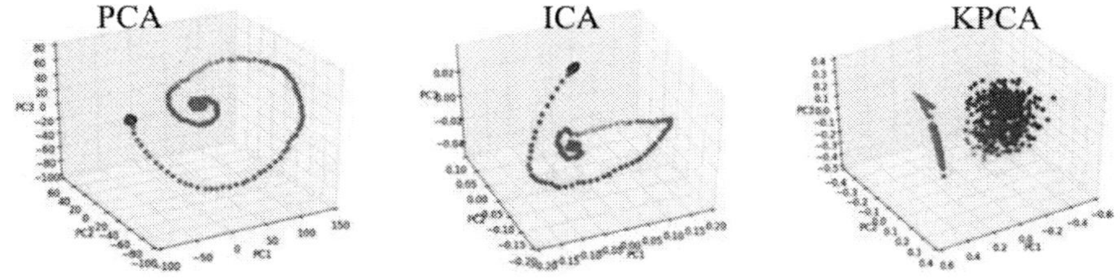

- **Different approaches** to DR achieve different representations of the data
- Manifold learning techniques: additional parameters and computational cost

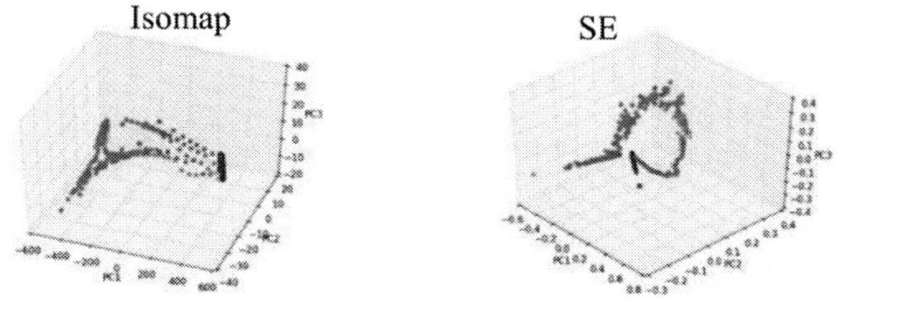

6

Data Clustering

- Divide data into groups that correspond to process behavior
- Steady state operations correspond to areas of high data density we can find with clustering algorithms
- Techniques:
  - Conventional: k-means: *k* centroids assigned to data

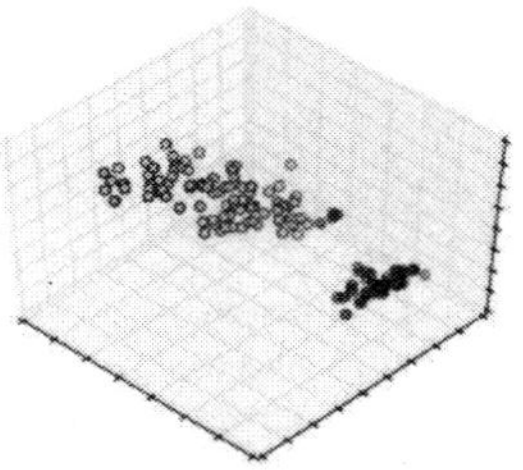

k-means converged at local minimum

Density-Based Approaches:

Basic idea: Clusters are dense regions in the data space, separated by regions of lower object density

- A cluster is defined as a maximal set of density connected points
- Discovers clusters of arbitrary shape
- Methods: DBSCAN-HDBSCAN

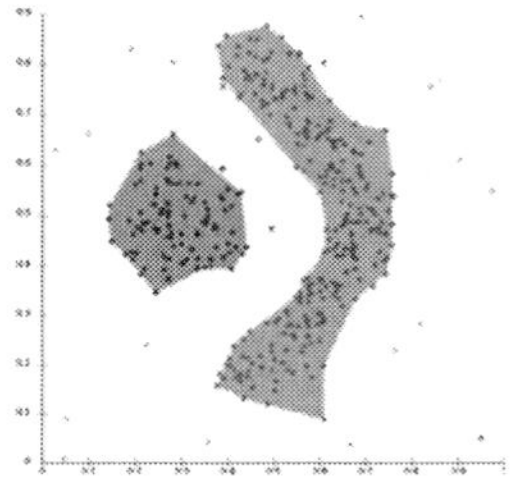

DBSCAN

Clustering Evaluation

Clustering evaluation metrics

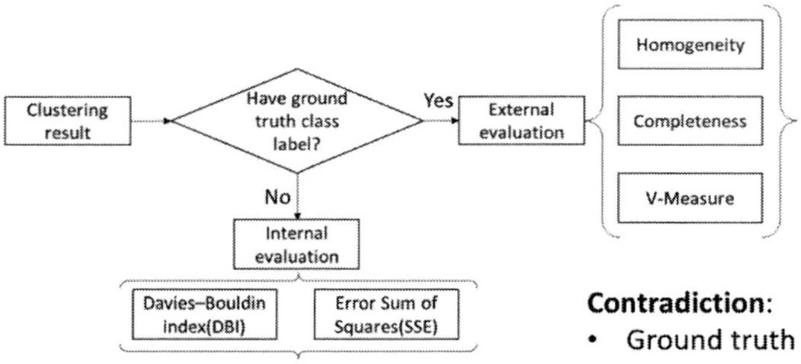

Contradiction:
- Ground truth class labels do not always exist
- Internal evaluation may not imply the objective knowledge

Solution:
- Cross validating clustering results by using alternative methods (e.g. SOM)

Self Organizing Maps (SOM)

- **Goal**: represent high dimensional data structure
- **Solution**: 2-dimensional approximation of data manifold that preserves topology of original data
- Visualizations show:
  - Relationships between clusters of data
  - Relationships between variables
- A data vector's *best matching unit* (BMU) is the map vector most similar to a data vector for classification
  - Used to show location of data on the map

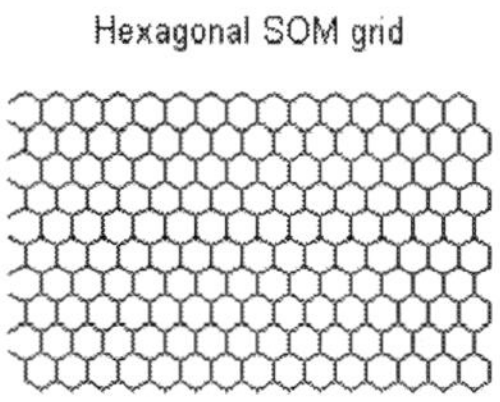

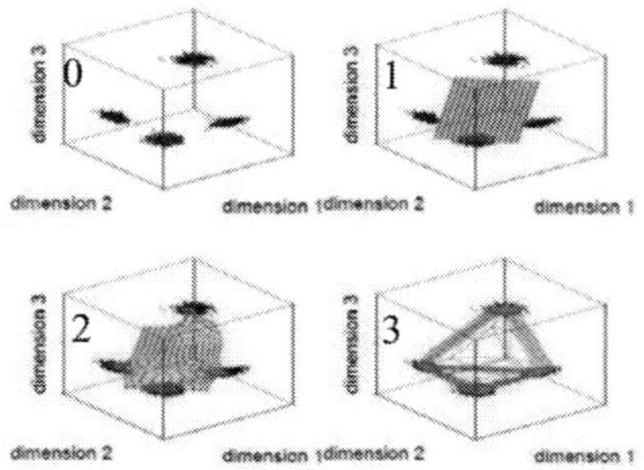

Self Organizing Maps (SOM)

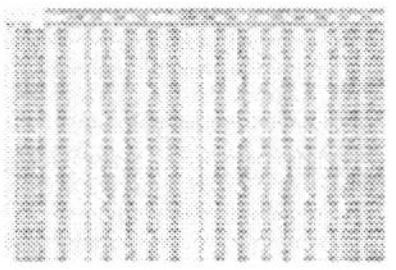

Units training

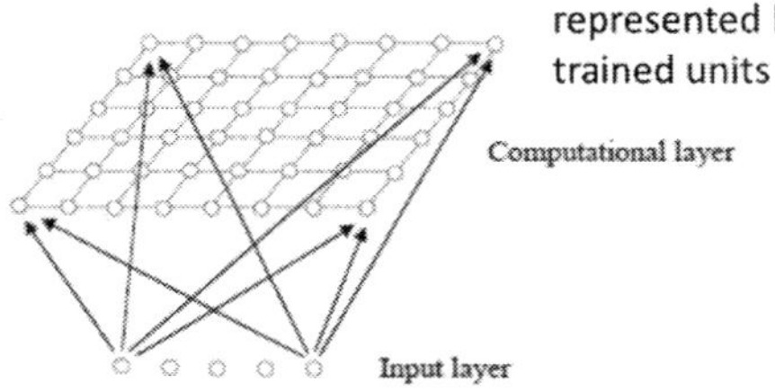

Original data is represented by trained units

Distance matrix between units pairs

Similar units have a low pair-wise distance (colored as dark blue). Clusters can be visualized.

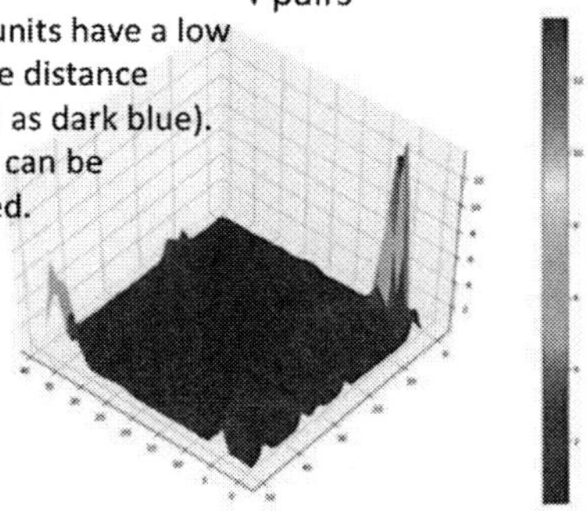

3D to 2D projection

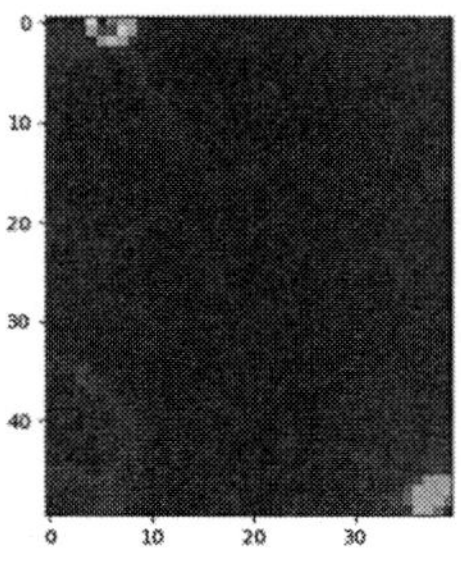

Explanation of clustered data

- Only clustering process data may not be enough to gain a complete understanding of the process

- Formation of clusters requires explanation
  - **Quantitative information** should be provided to explain the difference between pair-wise clusters.
  - Find combination of the **most contributing variables** to the difference between clusters.

- **A novel approach based on subspace greedy search is proposed**

The approach seeks for the combination of variables (subspace) which has the greatest contribution to the discriminant between different clusters(explanatory subspace).

11

Algorithm Demonstration

Comparing a data vector consisting of seven variables in cluster A to cluster B:

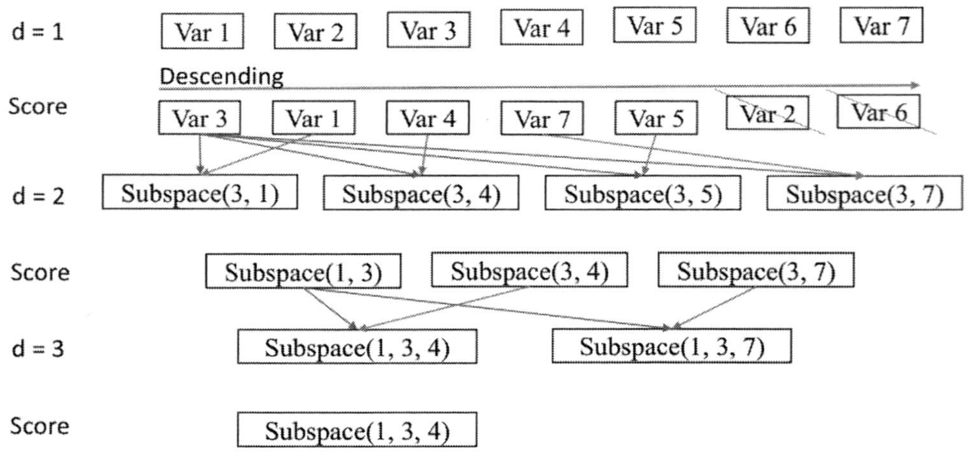

12

Remarks on algorithm

- Score function:
  - k-nearest-neighbor (kNN) distance can be robust to the nonlinearity of dataset.
- Computational cost:
  - Computational cost of a brute-force search is $O(2^d)$ for one data vector.
  - Greedy search algorithm is implemented to reduce the computational cost.
  - In each iteration, the greedy searching dramatically reduces the number of combinations.
- Ranking:
  - Occurrence frequency in candidates lists from each iteration can naturally rank their contribution.
- Data sampling:
  - For large dataset, data sampling can accelerate the computational speed for k-nearest-neighbor searching.

13

FASTMAN: Automated data mining tool

- Provides an **environment for data mining**: prototype different methods, relate clusters to time series plots, allows easy visualization
- Data from **different formats** can be considered

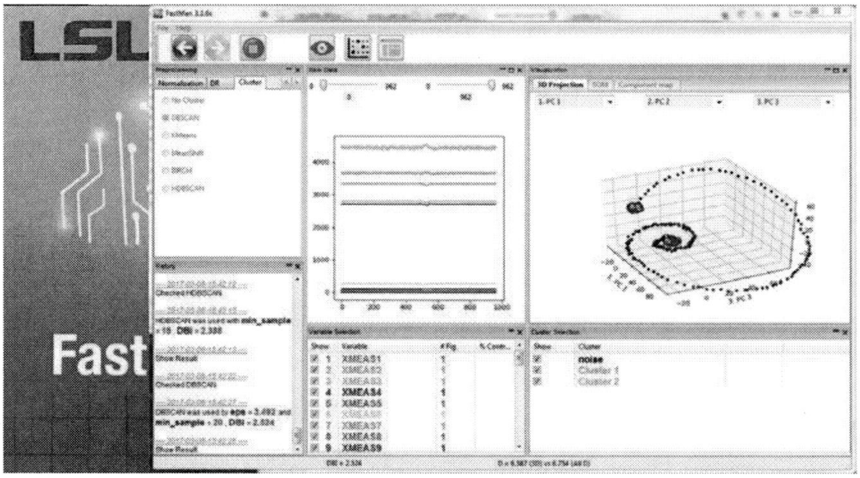

Implemented in the Python environment: free and open source software, supporting multiple programming paradigms such as object-oriented and functional programming styles.

Pyrolysis Reactor Process

- The pyrolysis reactor is used to crack heavier hydrocarbons to lower molecular weight hydrocarbon.

- **Six coils** of naphtha are mixed with steam and cracking in the furnace at 1000◦C.
- **27 measurements include**:
  - 6 HC flow rates
  - 6 steam flow rates
  - 6 crossover temperatures
  - 6 cracking temperature
  - 1 total HC flow
  - 1 overall outlet temperature
  - 1 furnace temperature
- **Operating cycle is about 3 months.**
- **Normal operation is drifting due to coking in coils.**

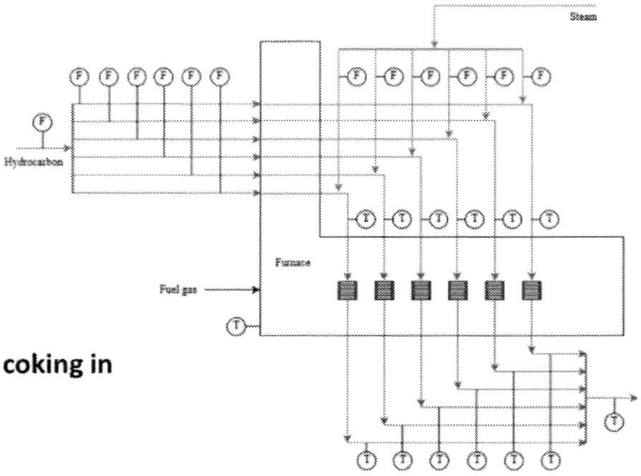

15

Pyrolysis Reactor Process

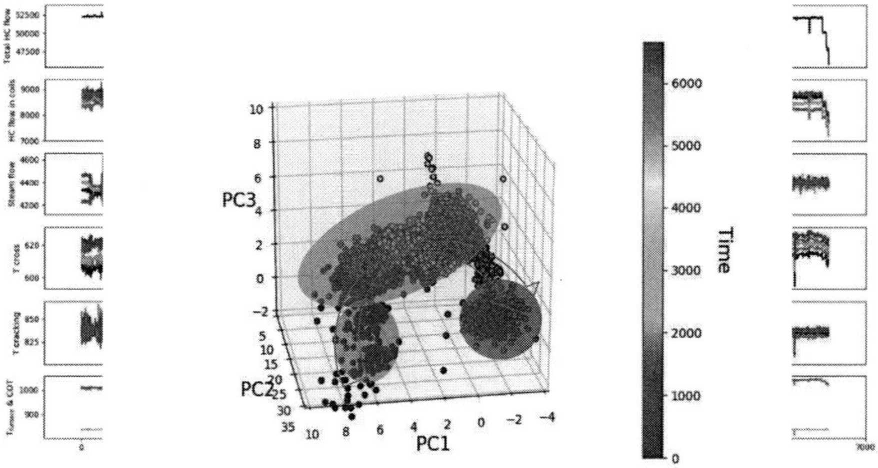

16

Application on Pyrolysis Reactor

- DBSCAN on process data

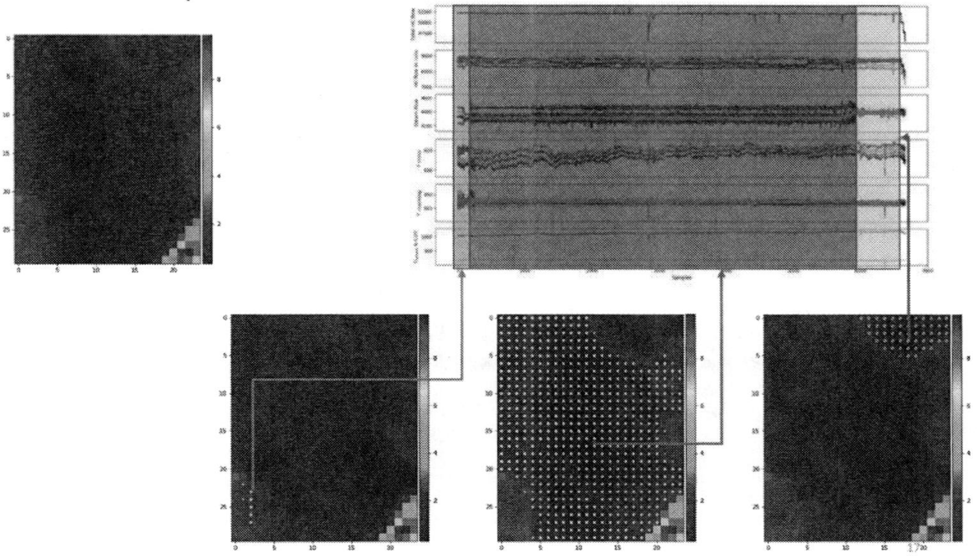

17

Application on Pyrolysis Reactor

- Explaining following two patterns:

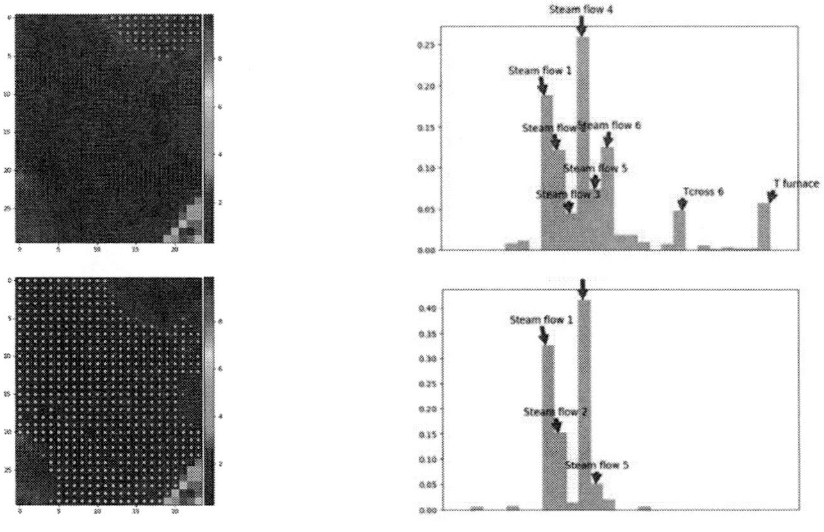

18

Application on Pyrolysis Reactor

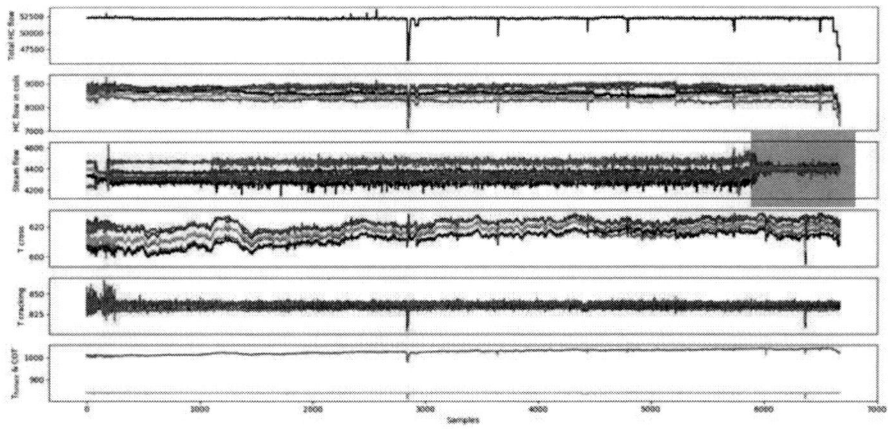

19

FastMan: Application on Pyrolysis Reactor

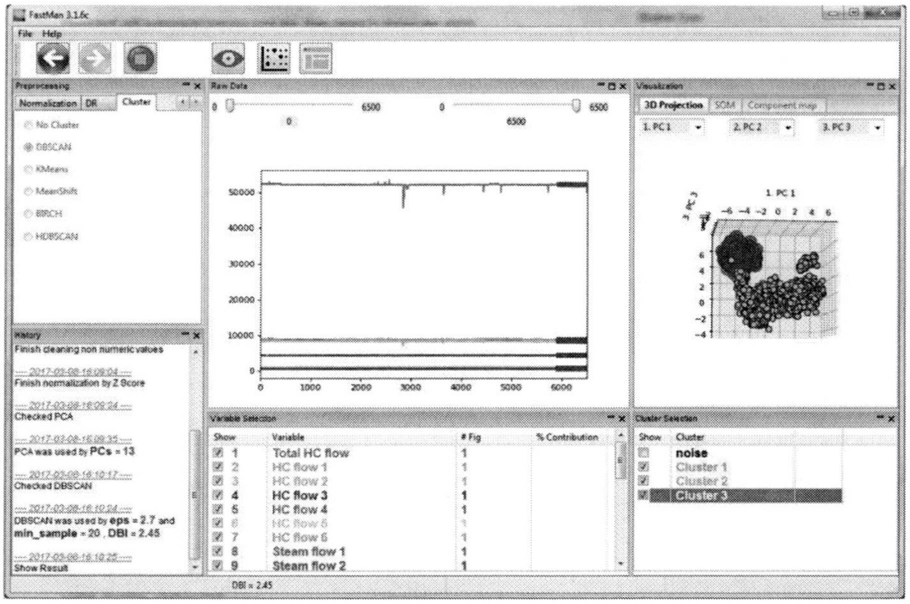

FastMan: Application on Pyrolysis Reactor

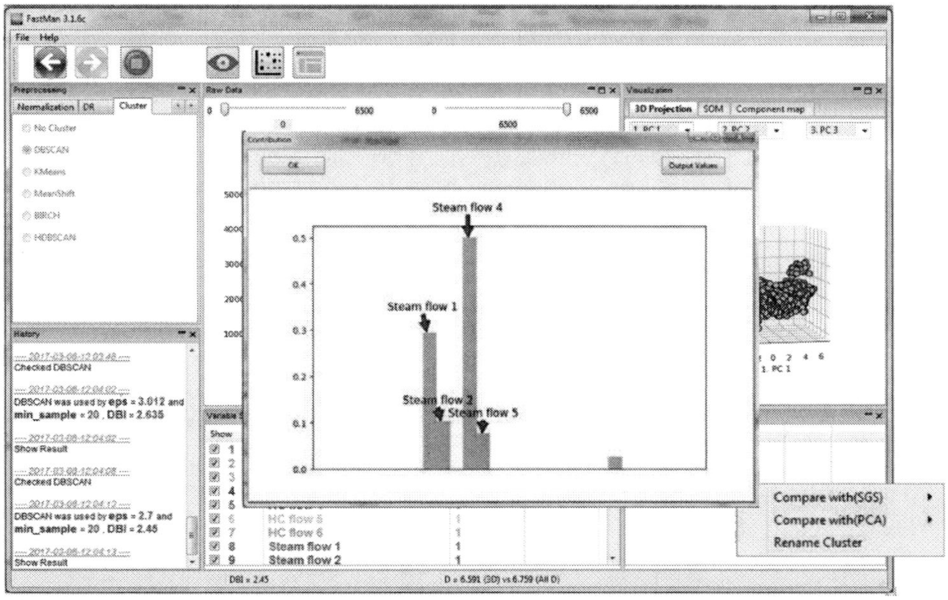

FastMan: Application on Pyrolysis Reactor

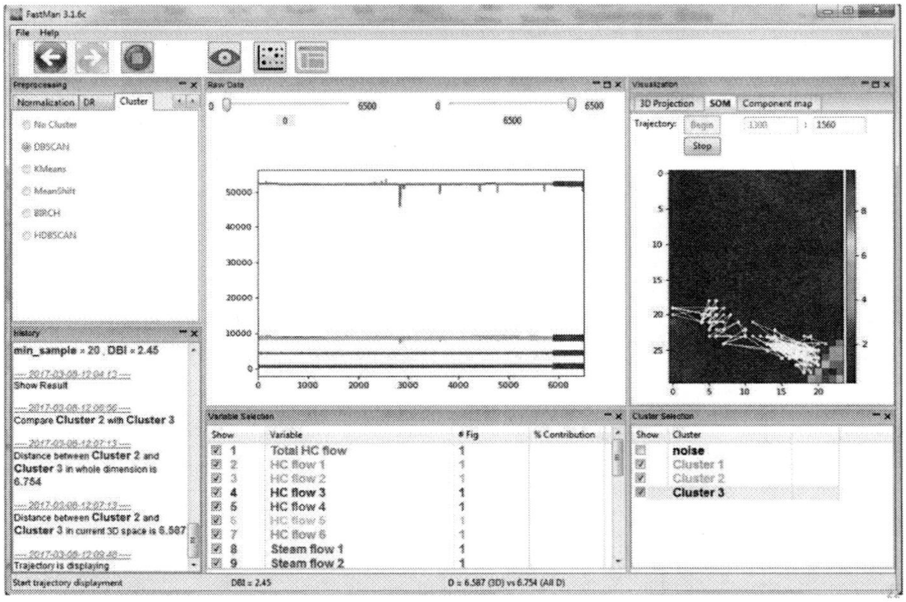

Conclusion

- We have studied the potential for **data clustering and unsupervised learning** to automatically separate data into groups significant to abnormal event detection and quality monitoring.
- **Applied** to clustering data from the benchmark Tennessee Eastman Process and from industrial scale Pyrolysis reactor
- **Extensive process knowledge** can be extracted from offline data mining approach
- **Cross validation** via SOM can ensure the clustering quality
- **Subspace greedy search** was proposed to reveal internal knowledge from clustered process data.
- **An integrated data-mining tool**, *FastMan* was developed to accelerate the offline data analysis.
  - Implemented in the Phyton environment providing full functionality and connectivity

23

Acknowledgement

Questions?

24

479996 Statistics for Big Data

Abstract:

After noting the relative absence of statisticians from the community of practice engaged with big data, we explain what big data is, how it's done, and who's working with it. Statisticians have much more to contribute in both the intellectual vitality and the practical utility of big data. At the same time, big data challenges statisticians to move out of some familiar habits to engage less structured problems, to become more comfortable with ambiguity, and to engage computer scientists in a more fruitful discussion of what the various parties can bring to this new mode of investigation. In this talk, we propose some potential directions for future research.

This talk is mainly based upon a 2014 joint paper with Dr. John Jordan.

481283 Big Data Analytics Applied to Semiconductor Manufacturing

Abstract:

Introduction

As "big data" requirements and capabilities in data volumes, velocity (rates), veracity (quality), variety (merging) and value (analytics) increase rapidly, semiconductor manufacturers are faced with a need for new approaches to data management and use across the fab [1]. During semiconductor manufacturing a silicon wafer is transformed into a grid of silicon chips in a fabrication facility or "fab". Each chip or "die" has layers of complex circuitry that must be controlled to the nanometer or even molecular level. The semiconductor manufacturing process usually has hundreds of these precision control steps. The dynamic control of these steps in this industry is facilitated with a set of capabilities collectively termed "Advanced Process Control" or APC [2]. Over the past two decades the APC family of capabilities has included dynamic model-based process control, termed "run-to-run" or R2R control [2,3], and Fault Detection and Classification or "FDC" (also referred to as excursion detection and failure mode determination) [2,4]. More recently the APC family of capabilities has been expanded to include predictive capabilities such as Predictive Maintenance (PdM), Virtual Metrology (VM) and even end-of-line yield prediction and yield excursion root cause analysis. The expanded set of APC capabilities represents the primary area where big data solutions are generating significant benefits in semiconductor manufacturing.

Big Data and Semiconductor Manufacturing Advance Process Control

Big data advancements have facilitated improvement in all of the APC capability types. The following are two examples where significant return on investment (ROI) is being achieved.

Predictive Maintenance: Predictive Maintenance, PdM, also referred to Predictive Preventative Maintenance, PPM, or a component of Prognostics and Health Management (PHM) is receiving considerable attention in semiconductor manufacturing because it is focused on reducing unscheduled downtime, which is a major source of quality loss and cost in manufacturing [5]. PdM is the technology of utilizing process and equipment state information to predict when a tool or a particular component in a tool might need maintenance, and then utilizing this prediction as information to improve maintenance procedures [2]. This could mean predicting and avoiding unplanned downtimes and/or relaxing un-planned downtime schedules by replacing schedules with predictions. In semiconductor manufacturing some PdM solutions are relatively straight-forward extensions of FDC, however most involve extensive data mining and analysis capabilities to develop models that not only predict the future occurrence of an equipment or component failure, but, as illustrated in Figure1, also indicate the time horizon for remaining useful as well as a confidence and range of the prediction. These PdM prediction events are linked into maintenance, inventory, scheduling and manufacturing execution systems to optimize uptime, cost, productivity and quality [5].

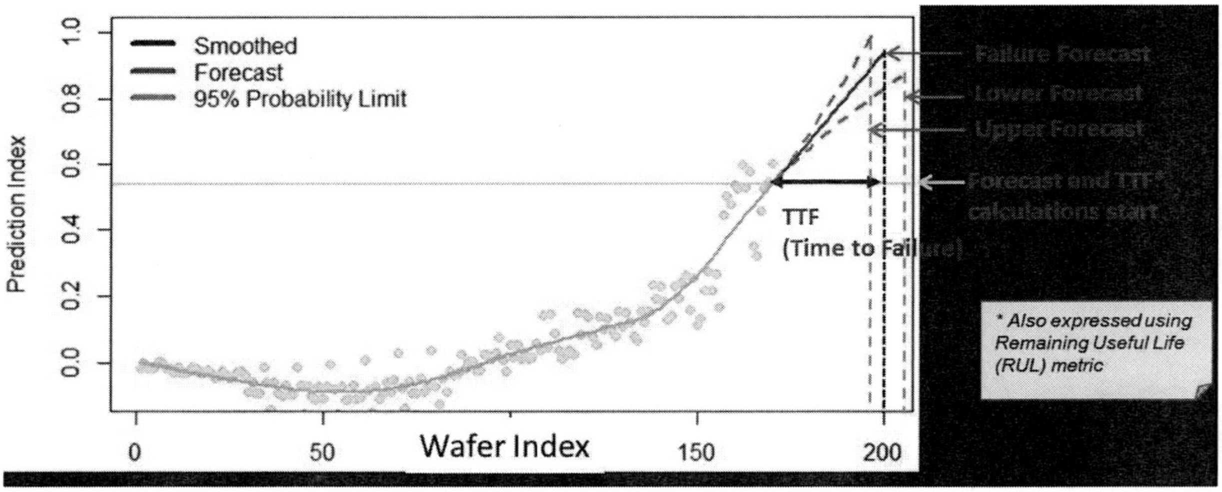

Figure 1: Illustration that PdM solutions require not only the actual prediction, but also a prediction horizon (Time to Failure-TTF) and an indication of prediction quality (provide here as a 95% probability range).

Next Generation FDC: Fault Detection (FD) utilizes real-time trace data collection of key equipment and process variables, along with process output quality information and advanced analytics to determine when an excursion has occurred. Fault Classification (FC) then tries to relate the excursion to a cause or classification. FDC is particularly challenging in semiconductor manufacturing because the processes complex and dynamic to the point that purely empirical solutions generally are ineffective. In other words equipment and process subject matter expertise (SME) must be incorporated into the development and maintenance of these models. As a result the FD model development process is usually highly manual and univariate leading to tens of thousands of models and millions of limits to monitor and manage across the fab. With the advent of big data capabilities the model and limits building and management can be semi-automated and optimized leveraging big data analytics advancements while retaining the ability to incorporate SME in solutions. New FD enhancements that are proving effective include semi-automated trace partitioning, feature selection, and merging in of process quality data to support feature ranking and limits optimization through receive-operating-characteristic (ROM) analysis. Interestingly many of the same algorithms leveraged for PdM can be applied here especially for ranking and optimization. The primary benefits over traditional FD are significantly reduced time for model setup and maintenance, and reduced false and missed alarms [6].

Industry Requirements for Big Data Analytics

The choice of analytics being leveraged to provide enhanced APC capabilities in semiconductor manufacturing reflects the complexity of semiconductor manufacturing processes in terms of the need for precision on 3 analytics response, the dynamics of the processes, and the need for incorporation of (human) SME into the analysis. As an example, typical multivariate supervised techniques such as Partial Least Squares (PLS), and Support Vector Regression (SVR) are often leveraged for predictive capabilities, however clustering techniques are also needed to help separate out the prediction event types, model adaptation techniques are needed to track the process dynamics, phenomenological model elements are often needed to improve empirically-tuned model accuracy, and the entire model building and maintenance process must be structured to support SME involvement. Additionally predictions must include an indication of prediction accuracy so that decision making can be optimized based on an understanding of the potential for false or missed alarms and the associated financial impact. All of these techniques benefit from the increased data volume and improved data quality for model building, however volume and quality of data to support robust models is still the primary impediment to the widespread success of prediction capabilities in the industry. Thus while many other manufacturing environments might turn to more general big data techniques, such as deep learning, for generalized analysis solutions, the semiconductor industry is currently focused on more guided techniques that leverage an understanding of the process, equipment and data management domain knowledge.

Summary

As the semiconductor industry advances into the era of smart manufacturing, specific requirements must be met to address the challenges of increasing manufacturing complexity. As shown in Figure 2, the big data infrastructure will enable a number of capabilities from manufacturing operations optimization, through supply chain network integration, and the leveraging of the simulation space and the concept of a "digital twin" for continuous fab optimization [7]. APC capabilities continue to be improved in order to optimize individual processing with respect to maximized quality and throughput and minimized cost. These capabilities include control, diagnostics, and more recently, predictive solutions. They leverage big data opportunities in analytics that allow for the incorporation of process, equipment and data expertise across the entire solution development and maintenance life cycle. It is expected that this approach will continue as the industry evolves from reactive to predictive to prescriptive computational analysis.

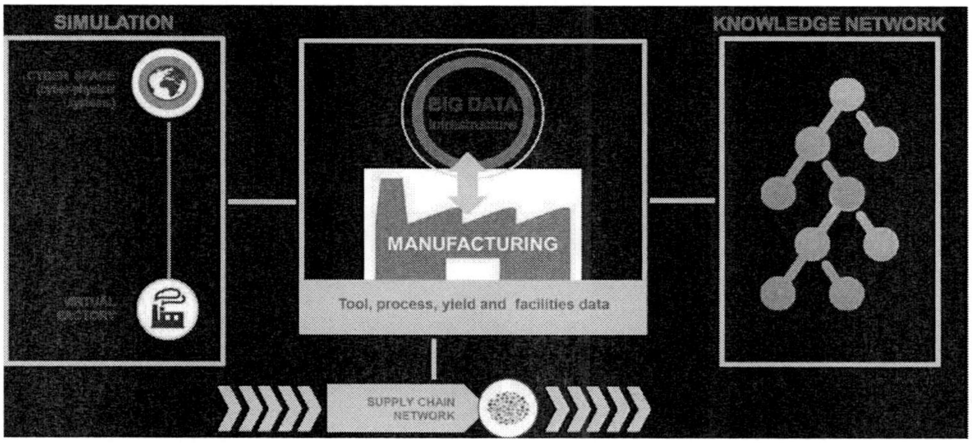

Figure 2: The Fab of the future requires integration of equipment, expertise, manufacturing capability and computational data analysis

References:

[1] J. Moyne, J. Samantaray and M. Armacost "Big Data Capabilities Applied to Semiconductor Manufacturing Advanced Process Control," IEEE Transactions on Semiconductor Manufacturing, Vol. 29, No. 4, November 2016, pp. 283-291.

[2] 2015 International Technology Roadmap for Semiconductors (ITRS): Factory Integration Chapter. Available: www.itrs2.net.

[3] *Run-to-Run Control in Semiconductor Manufacturing*, J. Moyne, E. Del Castillo, and A. Hurwitz, CRC Press, November 2000.

[4] T. Lee and C. O. Kim, "Statistical Comparison of Fault Detection Models for Semiconductor Manufacturing Processes," in *IEEE Transactions on Semiconductor Manufacturing*, vol. 28, no. 1, pp. 80-91, Feb. 2015.

[5] J. Iskandar, J. Moyne, K. Subrahmanyam, P. Hawkins and M. Armacost "Predictive Maintenance in Semiconductor Manufacturing: Moving to Fab-Wide Solutions," *Proceedings of the 26th Annual Advanced Semiconductor Manufacturing Conference (ASMC 2015)*, Saratoga Springs, New York, May 2015.

[6] J. Moyne, B. Schulze, J. Iskandar and M. Armacost "Next Generation Advanced Process Control: Leveraging Big Data and Prediction," *Proceedings of the 26th Annual Advanced Semiconductor Manufacturing Conference (ASMC 2016)*, Saratoga Springs, New York, May 2016.

[7] J. Moyne, "Emerging trends in IC manufacturing analytics and decision making," APC Conference XXVIII, Phoenix, AZ, October 2016. Available: www.apcconference.com.

480087 Real-Time Performance Monitoring and Diagnostics of Rod Pump System for Oil Elevation Using Streaming Dynamometer Card Data

Abstract:

Rod pump systems are the most commonly used mechanism for oil lift in the world. The ability to monitor the rod pump operation and diagnose any impending problems with the pump and/or the well is therefore of primary interest to oil field operation. Downhole dynamometer cards (obtained by plotting the measurements of the rod displacement vs. the computed load on the pump, originally developed by Walton E. Gilbert) are viewed as the main information source for performance monitoring and diagnostics of the rod pump and the well today.

Methods to deduce the operating condition of the well and the pump from downhole dynamometer card data are reported both by academia and industry. The use of downhole dynamometer cards are at the heart of many commercial software/hardware products that target performance monitoring and diagnostics of rod pump systems in oil fields around the world.

In this paper, we present an algorithm for automated performance monitoring and diagnostics of the well and pump operating conditions where the streaming card data is processed in real-time. The proposed algorithm uses a hybrid classification approach where both labeled and unlabeled card data would be consumed to generate insight for the operators. The key objective here is to enable the detection of an anomaly in the operation by maximally utilizing prior knowledge while enabling automatic capture of the unique characteristics of a well/rod pump unit solely based on the live operation data.

480732 Challenges in Applying Data Analysis to Industrial Processing: Inverse Problem Viewpoint

Abstract:

Production data analysis is often a regular part of the evaluation and troubleshooting for operations. This is especially true when looking at issues of quality, conversion rates, production rates, or rates of degradation in the process. Ideally, it would be valuable to understand what is actually happening inside the production units. In particular, production problems would be easier to address if it could be determined where and when in the process significant changes are occurring. This can be considered an inverse problem of determining the process configuration that gives the measurements seen.

The underlying physics can provide insights to data analysis and is probably necessary to solve the "Inverse Problem". For example, the physics of a fixed bed reactor will produce relationships about the flows, pressure drop, temperature, heat generated (absorb), and stream composition changes. Applying these physics in an industrial setting will require numerous assumptions about the process. There may be competing assumptions that need to be considered. For the reactor example, these assumptions will be capture in parameters that designate the flow resistance and reaction kinetics. The competing assumptions will generate different model structures and possibly different types of parameters. The analysis of the production data will drive refinements of these parameters.

This presentation examines the challenges and tools for analyzing production data and for determining the appropriate physic driven models that can capture the behavior of the production units. This will include approaches for doing time series analysis, uncertainty quantification, principle component analysis, machine learning, objective function determination and data structure analysis. In particular, we will look at how these tools can provide different insights into the underlying production behavior.

Investigating Polymeric Compositions

via Clustering of Production Analytical Characterization Data

Alix Schmidt, Jennifer Wightman
3/29/17

Agenda

Silicone Resin Overview

Introduction to Data

Clustering

Interpreting Cluster Results

Conclusions

Silicone Overview

From quartz to highly specialized materials, >7,000 products

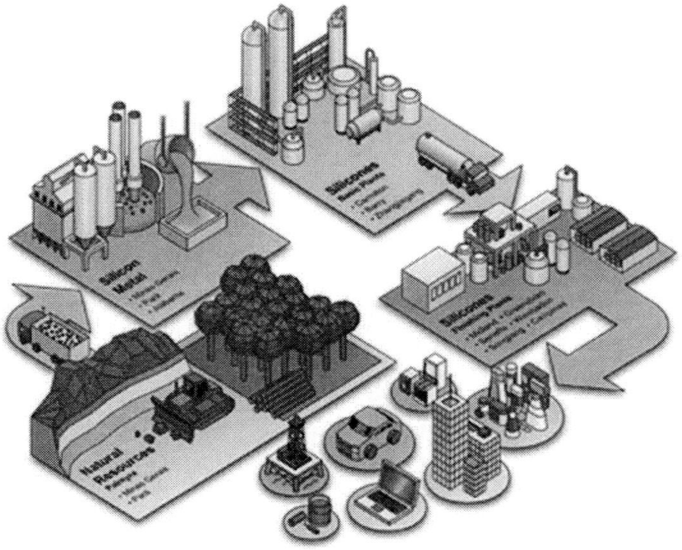

3

Silicone Resin Production

Nomenclature: M = $R_3SiO_{1/2}$, D = $R_2SiO_{2/2}$, T = $RSiO_{3/2}$, and Q = $SiO_{4/2}$

Hydrolysis and Condensation of Si monomers to grow MW:

- Chlorosilanes, Alkoxysilanes, Silicates

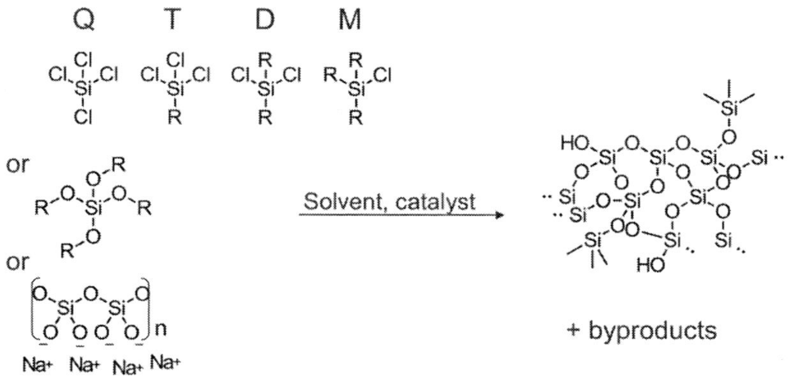

4

Silicone Resin Types

$M = R_3SiO_{1/2}$, $D = R_2SiO_{2/2}$, $T = RSiO_{3/2}$, and $Q = SiO_{4/2}$

MQ

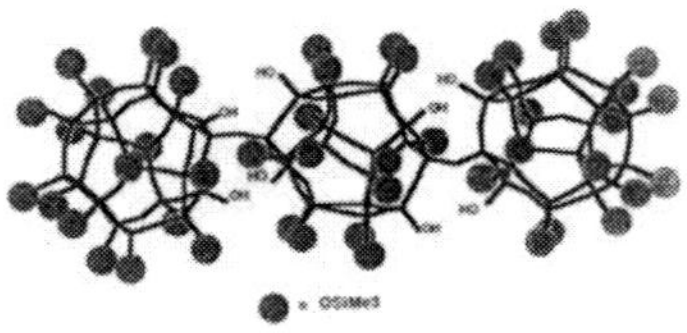

$MT^{Ph}D^{Me}$ [1]

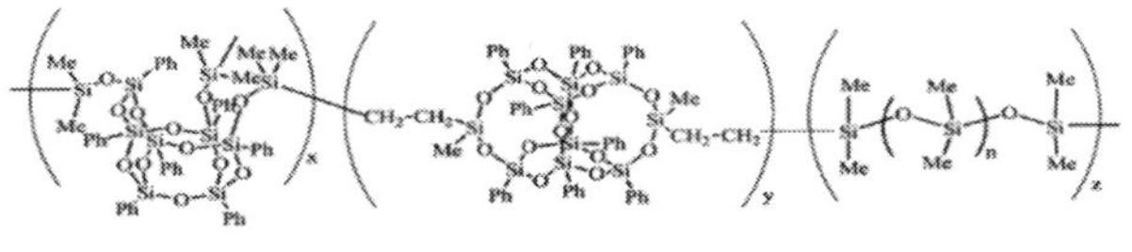

T

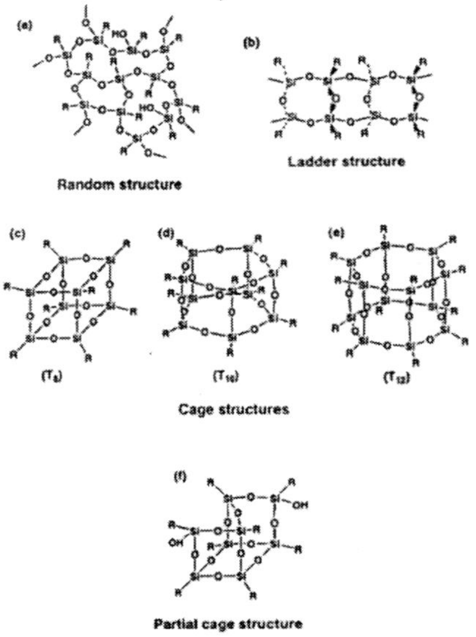

Endless possibilities for structured, amorphous, and hybrid materials....

[1] M. Aminuzzaman, A. Watanabe and T. Miyashita, *Journal of Materials Chemistry*, vol. 18, no. 42, pp. 5092-5097, 15 August 2008.

5

Resin Characterization Challenges

Molecular characterization of these complex structures is not straightforward

- Gel Permeation Chromatography (GPC) for size
- Viscometry/rheology for size and flow properties
- Gas chromatography for impurities/byproducts
- Titrations for impurities
- Optical properties for structure/impurities
- FTIR for structure

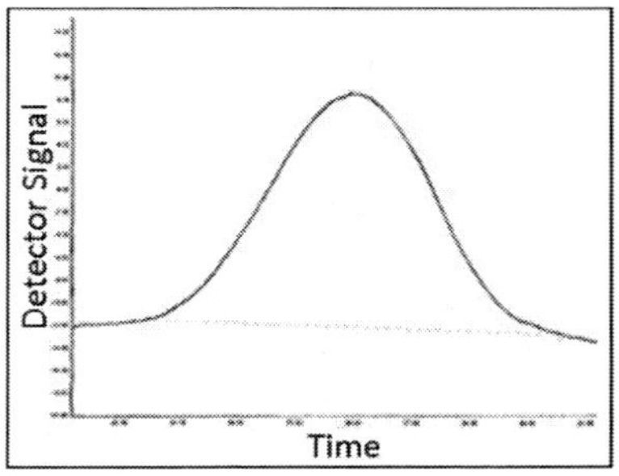

6

What is meant by "Product Quality"?

Options for quality control:

- Many univariate specifications
- Multivariate control charting
- Advanced modeling of conditions

OR

Can we find a multivariate classifier to define a specific *structure or composition,* rather than the physical properties?

Hypothesis:

Different process conditions produce different resin structures, which inherently have different feature patterns

7

What can we do with historical data?
▸ Unsupervised Learning

The silicone resin structure exploration is an *unsupervised* learning problem → we have all x's and no y's

- Not a predictive model for product quality (yet)
- Mining the data to identify patterns that relate to our hypothesis
- Data analytics as a tool in the scientific method

8

Introduction to the Data

14 Analytical Characterization Results for 770 Batches

Purity:

- Non-volatile content (NVC)
- Residual Impurity A
- Residual Impurity B
- Residual chloride

Physical properties :

- Viscosity
- Refractive index (RI)
- "Cut viscosity" at a fixed NVC

Extent of the condensation reaction:

- Silanol (SiOH) concentration
- Molecular size from gel permeation chromatography (GPC):
    - Mn, Mw, Mz, and polydispersity (PD) from GPC Method 1
    - Mw from GPC Method 2
    - A proprietary molecular weight descriptor, "HMS"

9

Exploratory Data Analysis

Inspect distributions of the 14 variables

- Values range from 10^{-3} to 10^{7} → all were standardized to ~N(0,1)
- All very normally distributed except Impurity A and Mz

Viscosity | Mz

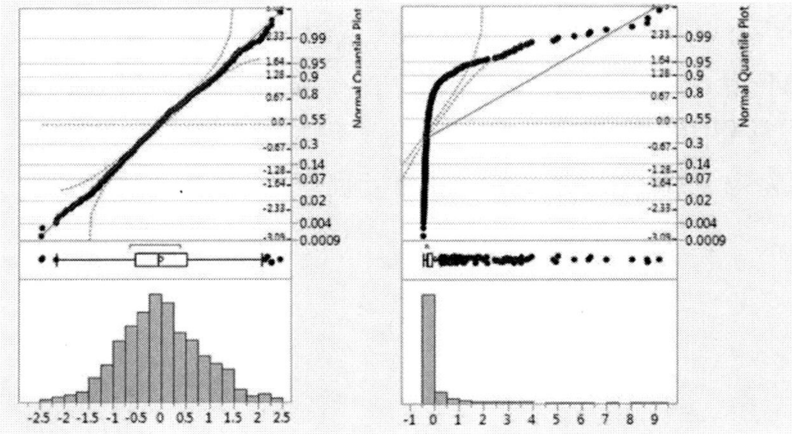

10

Cluster Analysis

Finding the best way to group the observations when you don't have the "right" answer

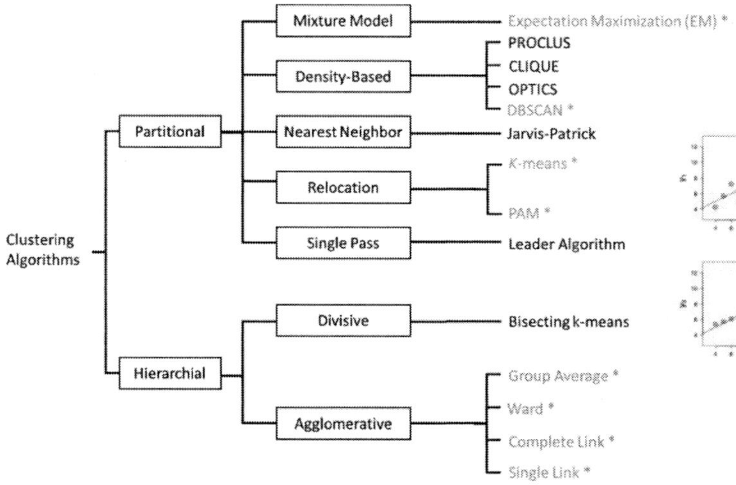

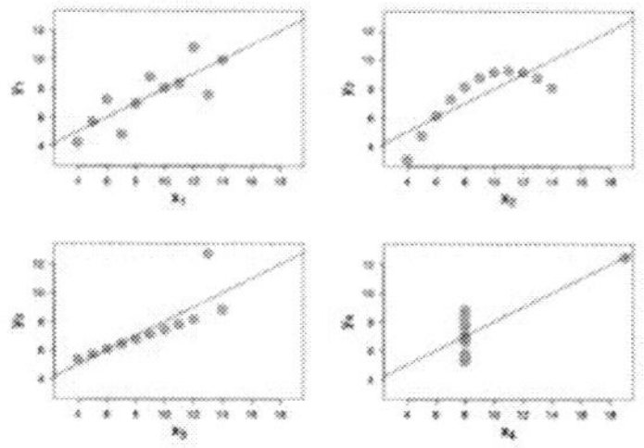

(left) [3] G. Downs and J. Barnard, "Clustering Methods and Their Uses in Computational Chemistry," in *Reviews in Computational Chemistry*, vol. 18, K. Lipkowitz and D. Boyd, Eds., John Wiley and Sons, Inc., 2002, pp. 1-40.
(right) [4] Anscombe, Francis J. (1973) Graphs in statistical analysis. American Statistician, 27, 17–21

13

Visualization of 14-Dimensional Clusters

Principal Components Analysis can help

No obvious cluster separation

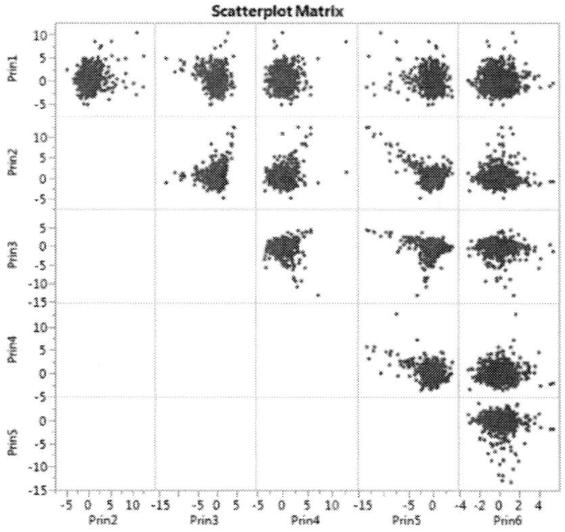

14

K-means and K-Medoids (PAM)

Short story: choose the number of clusters (*k*), assign every point to the closest cluster center, then iterate locating the center and reassigning points until the clusters are stable

Tuning parameters:

- What does cluster "center" mean?
- What does "closest" mean?
- How do you choose *k*?

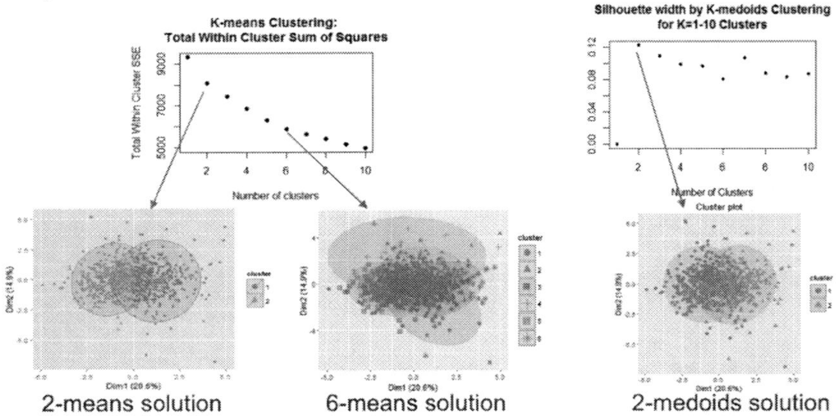

2-means solution 6-means solution 2-medoids solution

Density-Based (DBSCAN)

Short story: can identify clusters of complex shapes by detecting areas of low data density between groups

Tuning parameters:

- Minimum # of points to define a cluster vs. "noise"
- The max distance allowed

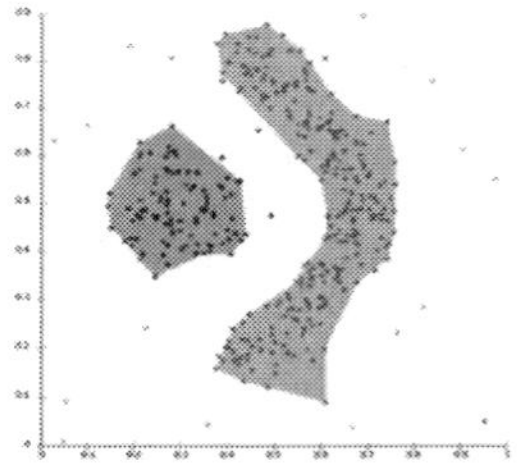

Example of DBSCAN clustering, drawn from Chire [6]

Results: Found 1 cluster with lots of noise

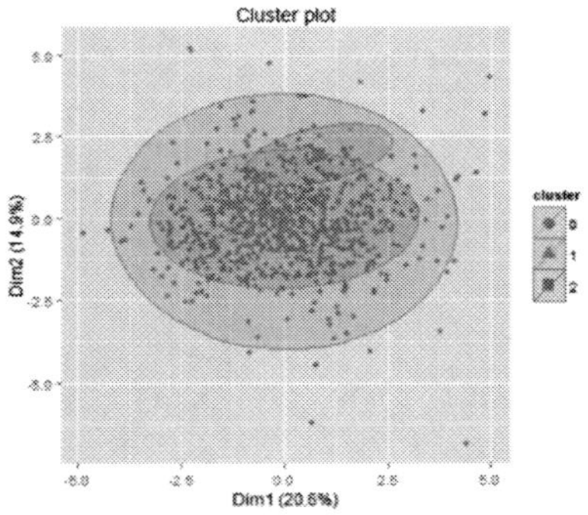

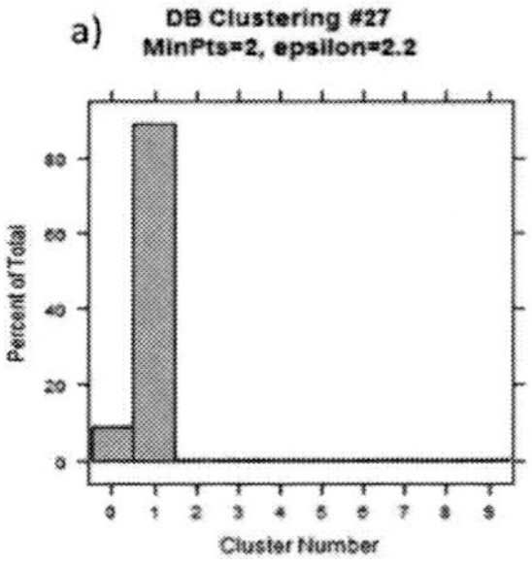

[6] Chire, "DBSCAN-density-data.svg", commons.Wikimedia.org, 2012, CC-BY-SA-3.0

Expectation Maximization (EM)

Short story: Estimate parameters for many mixtures of *k* multivariate normal distributions, then select the mixture and *k* which maximizes BIC

Tuning parameters: *k*

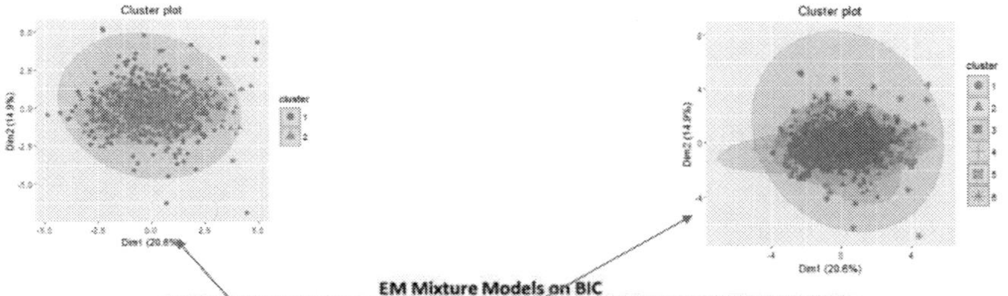

Resin data appears to be ellipsoidal

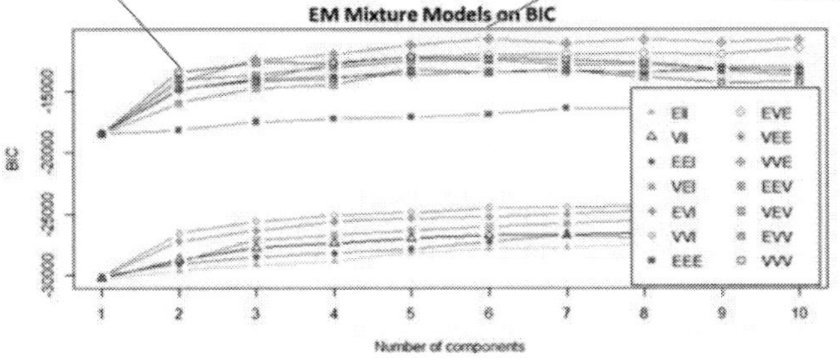

(limitations on possible shapes of multivariate normal distributions)

19

Hierarchical (Connectivity-Based)

Short story: Find the two closest points/clusters, combine them, then find the next closest, and the next, and the next… from N to 1 clusters

Tuning parameters – what does "closest" mean?

- Connection type
- Distance metric

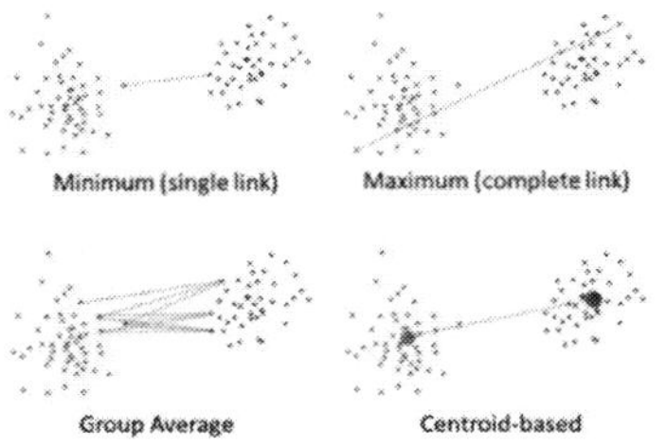

Results: Wide range of solutions based on connectivity!

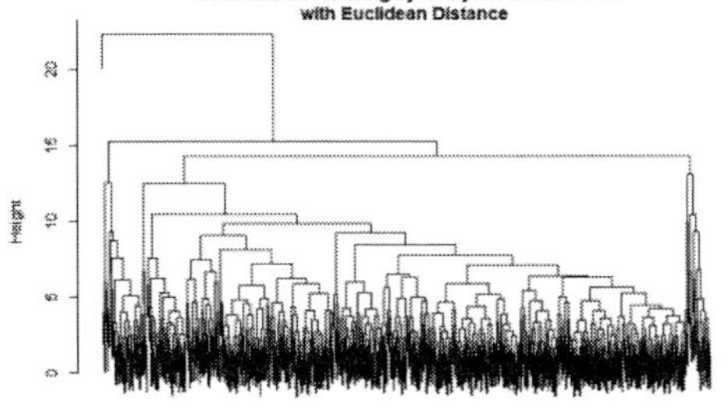

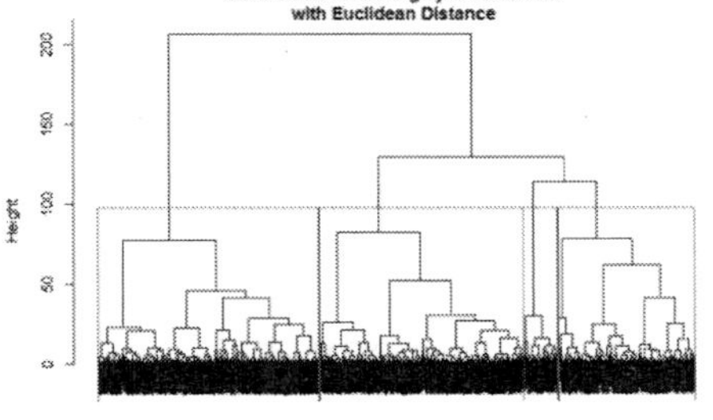

20

Which Results are Best?

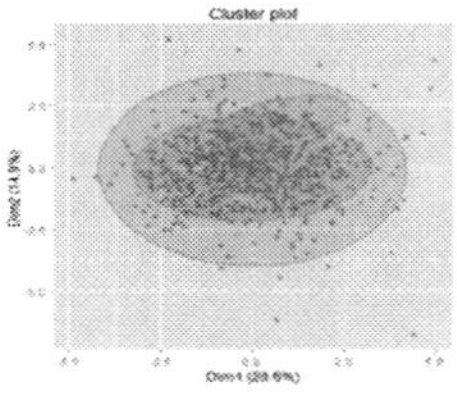

DBSCAN solution

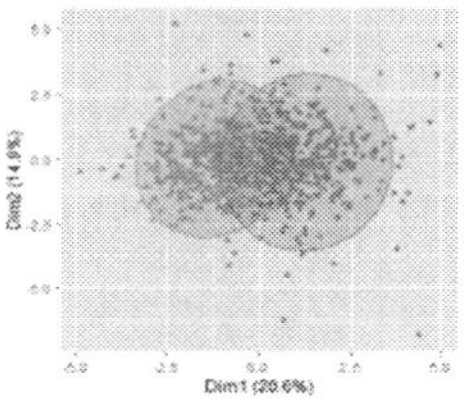

2-means solution

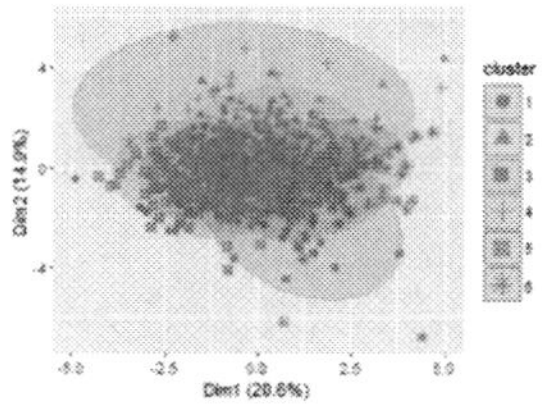

6-means solution

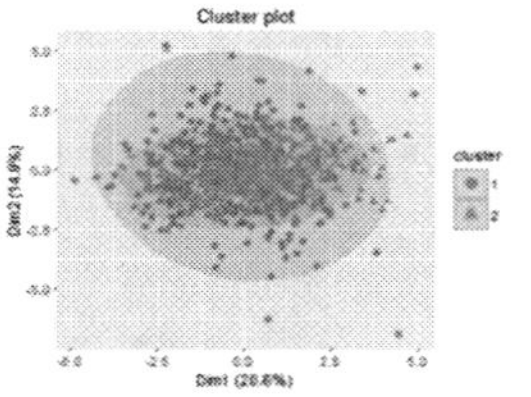

EM 2 solution

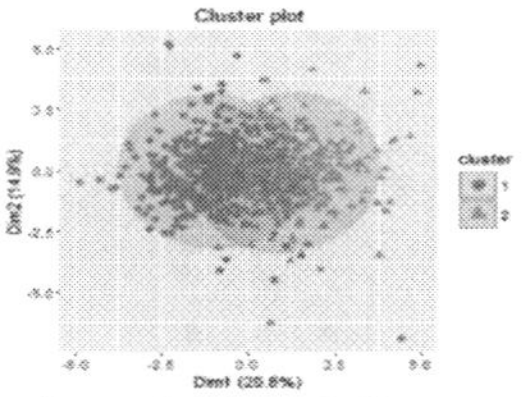

2-medoids solution

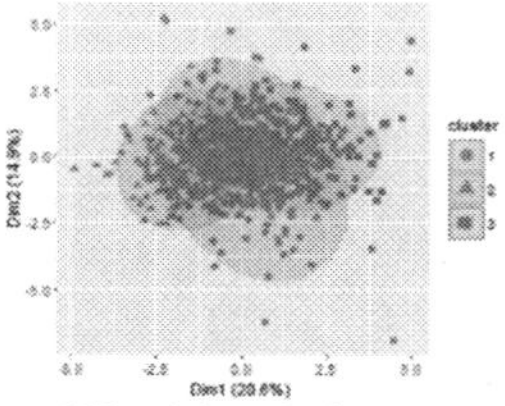

Ward 3 solution

21

Internal Cluster Validation Indices

Cluster validity indices (CVI) compare the within-cluster cohesiveness to the between-cluster separation

Challenges :

- Like R^2, some CVI necessarily improve with more clusters
- CVI ignore small clusters [8]
- Trouble comparing different size or density clusters [8]
- Many CVI penalize ellipsoidal clusters[9]

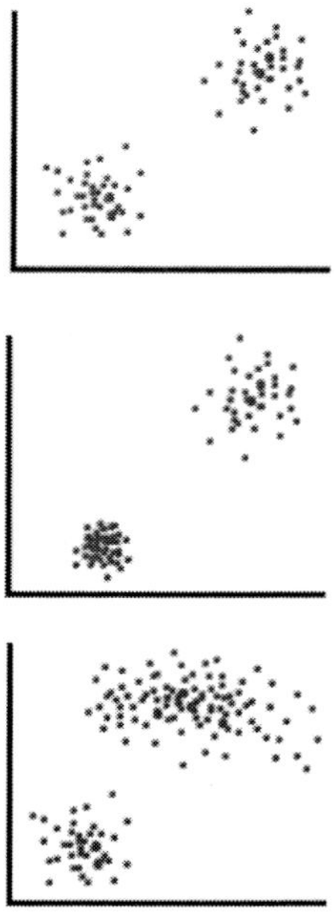

[8] K. Zalik and B. Zalik, *Pattern Recognition Letters,* vol. 32, pp. 221-234, 2011.
[9] O. Arbelaitz, I. Gurrutxaga, J. Muguerza, J. Perez and I. Perona, *Pattern Recognition,* vol. 46, pp. 243-256, 2013.

22

CVI Results

No one solution stands out

Clustering Method	# of clusters	Cohesion		Separation		Ratios, etc.		
		Max Diameter	Average WSSE	Min Separation	Average BSSE	Dunn	C-H	Average Silhouette Width
		↓*	↓*	↑*	↑*	↑*	↑	↑
K-means.2	2	21.9	4.1	0.8	4.8	0.035	126.6	0.14
DBSCAN.2	2	22.3	4.4	0.9	4.5	0.040	17.5	-0.07
PAM.2	2	22.3	4.2	0.8	4.7	0.034	107.3	0.12
EM.2	2	22.3	4.1	0.9	5.0	0.039	37.0	0.14
Ward.2	2	22.3	4.3	1.0	4.6	0.045	82.1	0.08
Ward.3	3	22.3	4.0	1.0	4.7	0.045	73.7	0.09
K-means.4	4	20.8	3.8	0.7	4.7	0.035	97.1	0.12
PAM.4	4	22.1	3.9	0.5	4.6	0.025	79.6	0.10
Ward.4	4	21.7	3.8	1.0	4.7	0.046	78.0	0.10
K-means.6	6	18.9	3.4	0.7	4.7	0.038	95.2	0.13
EM.6	6	22.3	3.9	0.6	4.6	0.025	19.3	-0.02
Ward.6	6	18.1	3.6	1.0	4.7	0.055	70.3	0.09
DBSCAN.8	8	22.3	4.0	2.3	6.6	0.103	6.2	0.04

Revisiting the Hypothesis

Hypothesis:
Different process conditions produce different resin structures, which inherently have different feature patterns

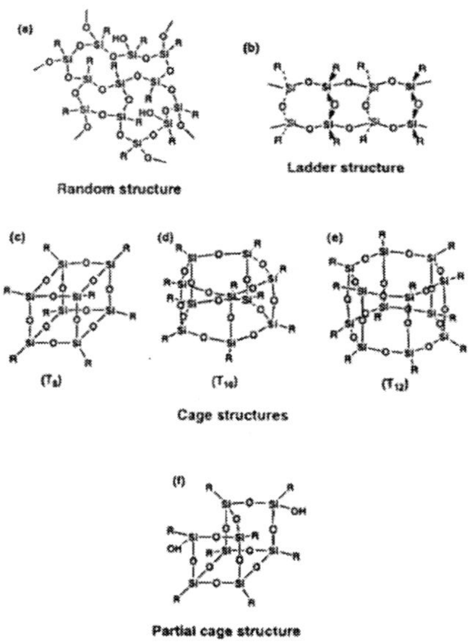

Findings:

There is no evidence to suggest distinct structures are formed, and/or the feature patterns of the compositions are not well resolved from process noise

Interpreting Cluster Results – Fundamental Perspective

The 2-means solution did have predictive power for downstream processing results

Work with the subject-matter experts to interpret the meaning of the clustering results

Focus on both what is different *and* what is the same

Keep in mind practical significance, not just statistical significance

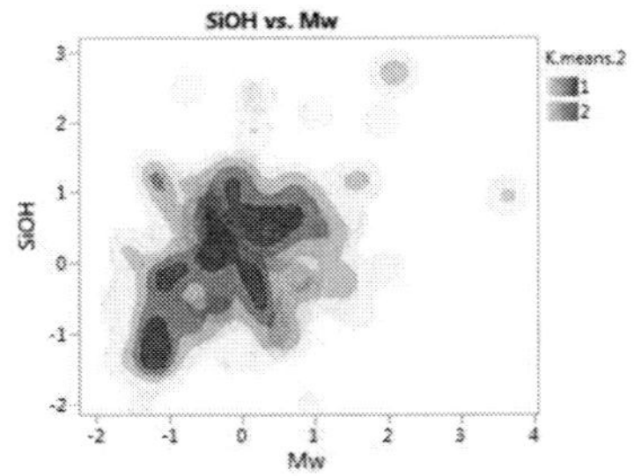

26

Conclusions

We strive to align our process control strategy to fundamental engineering and chemistry... not so easy with polymeric mixtures

Multivariate Analysis can help... but clustering has a lot of caveats

Even poorly separated clusters can still be useful predictors for product performance

30

Northwestern University

Acknowledgements

Dr. Jamie Riggs
Dr. Gary Wieber
Thomas Bekemeier
Kirk Opella
Robert Fosdick
Tim Morrison
Dr. James Mentele
Dr. Kurt Brandstadt

References

[1] M. Aminuzzaman, A. Watanabe and T. Miyashita,*Journal of Materials Chemistry,* vol. 18, no. 42, pp. 5092-5097, 15 August 2008.

[2] R. G. Lawson and P. C. Jurs, "New Index for Clustering Tendency and Its Application to Chemical Problems," *Journal of Chemical Information and Computer Science,* vol. 30, no. 1, pp. 36-41, 1990.

[3] G. Downs and J. Barnard, "Clustering Methods and Their Uses in Computational Chemistry," in *Reviews in Computational Chemistry*, vol. 18, K. Lipkowitz and D. Boyd, Eds., John Wiley and Sons, Inc., 2002, pp. 1-40.

[4] Anscombe, Francis J. (1973) Graphs in statistical analysis. American Statistician, 27, 17–21

[5] P.-N. Tan, M. Steinbach and V. Kumar, Introduction to Data Mining, Pearson, 2005

[6] Chire, "DBSCAN-density-data.svg", commons.wikimedia.org, 2012, CC-BY-SA-3.0

[7] T. Hastie, R. Tibshirani and J. Friedman, The Elements of Statistical Learning, 2nd ed., Springer, 2009.

[8] K. Zalik and B. Zalik, *Pattern Recognition Letters,* vol. 32, pp. 221-234, 2011.

[9] O. Arbelaitz, I. Gurrutxaga, J. Muguerza, J. Perez and I. Perona, *Pattern Recognition,* vol. 46, pp. 243-256, 2013.

R Libraries

- amap
- analogue
- caret
- cluster
- clustertend
- dbscan
- fpc
- ggplot2
- mclust
- NbClust
- Pvclust
- vegan

33

480673 Design of Dynamic Experiments

Abstract:

In this tutorial presentation we describe the first generalization of the Design of Experiments (DoE) methodology, the Design of Dynamic Experiments (DoDE). Its main and powerful characteristic is that it allows the experimental factors to be functions of time instead of constant values. This enables us to consider, for example, time variations in the reaction temperature or time variations in the feeding of a co-reactant. The power of this new methodology is that is offers a unique data-driven avenue for the optimization of processes for which a detailed knowledge-driven model is not at hand. The presentation will include several examples of batch processes including a Dow polymerization reactor and a pharmaceutical hydrogenation reaction. The productive of the Dow process was increased by 19%, shortening the batch time while producing the same amount of product with the desired quality. The selective and economics of the pharmaceutical hydrogenation was increased by 40% by operating the reactor with a decreasing temperature profile.

479921 Alamo

Abstract:

The ALAMO software has been developed in order to infer algebraic models from data obtained from simulations or experiments. ALAMO uses a small number of data points in order to learn models that are as accurate and as simple as possible. The approach relies on optimization techniques in order to: (1) build low-complexity models of systems from input/output data, (2) adaptively sample new simulation or experimental points, and (3) enforce physical constraints on the response function. This tutorial will illustrate ALAMO and its application in several domains.

480109 Plug-in Analytics Device Hardware and Applications for Rockwell Automation's Connected Enterprise and the Internet of Things

Abstract:

The abundance of sensor data, demand for lean and efficient plant operations, and added complexity of supply chains have evolved industrial requirements and expectations for advanced analytics applications, yet many proposed solutions to meet these new challenges require extensive engineering expertise or massive historical data.

Rockwell Automation has developed innovative, plug-in IoT solutions for analytics. These hardware-based applications require minimal engineering expertise by combining Control Logix technology with data-driven process modeling concepts. Further, these applications access data streams on the connected bus to learn and adapt, minimizing the need for any process historian.

We present a three pump application, which utilizes two of Rockwell's core plug-in analytics capabilities: automated stream-based modeling and advanced dispatch scheduling to meet anticipated demands.

481014 Model Reduction for Complex Systems Analysis

Abstract:

We present results from the application of the ALAMO methodology to the development of reduced-order models (surrogate models) for several case studies of interest to the design, analysis, control, and optimization of chemical processes. We begin by considering several simple systems, including flash calculations and steam tables. We then assess the capabilities of ALAMO to build simple and accurate models for an air separation unit. Air separation is an energy intensive process that involves heat integration and material recycles between columns. As a result, this system is often difficult to model. A high-fidelity model for an air separation unit was developed in gPROMS. ALAMO was utilized in order to reduce the complex simulation model to a reduced-order algebraic surrogate model. Extensive computational results are presented to compare the two models in terms of run time and relative accuracy.

481016 Feasibility Mapping and Optimization Using Big Data

Abstract:

With the advent of high-fidelity model-based computer simulations and advanced sensors, more and more data is made available for the determination of feasible operating region and the optimization of complex chemical processes. Specifically, data-driven methods can be advantageous for black-box process optimization, where the analytical forms of the objective function and constraints are either unknown or are difficult to formulate explicitly leading to lack of derivative information. However, the high-volume of data and the high-dimensionality of real industrial problems can introduce significant challenges for data-driven feasibility mapping and optimization. Current data-based approaches are mostly limited to small data sets or samples and perform well only if the problem has a few decision variables. To this end, we propose an efficient method for the approximation of convex or nonconvex, continuous or disjoint feasible regions of constrained black-box problems. Here, we approximate a feasible region as the region described by the convex hull of all feasible samples and their neighboring points while subtracting the infeasible samples and their neighboring points from the convex hull. To address high-dimensional problems, we propose a novel approach that transforms a multi-dimensional problem into a single-dimensional parametric problem over a newly introduced auxiliary variable [1]. In this presentation, we will introduce the convex-hull based feasibility mapping and single-dimensional projection methods using data obtained from high-fidelity process models and discuss the results for the intensification of an integrated carbon capture and conversion process to generate syngas utilizing power plant flue gas and natural gas within a single column.

References:

1. Bajaj, I., Hasan, M.M.F. A Novel Derivative-Free Optimization Method based on Single Dimension Projection. Accepted for publication in the Proceedings of FOCAPO/CPC 2017.

480933 Big Data Analytics Approaches Applied to Chemical, Energy and Semiconductor Industries

Abstract:

Big Data Analytics Approaches Applied to Chemical, Energy and Semiconductor Industries

480674 *Advances in Multivariate Analysis*

Abstract:

TBD

478620 Using Analytics for Process Safety

Abstract:

Managing plants flooded with thousands non-statistically based alarms daily creates a major challenge for plants wanting to minimize risk. Operators are dismissing non-critical alarms throughout the day, but how many of non-critical alarms does it take to make a critical alarm? 35? 75? 1,000? (is this death by 1,000 cuts?)

Using real-time analytic monitoring can create a scientific way of filtering the data to provide early warning detection, while reducing the noise from fixed set point alarms which may not be relevant for the current process. This presentation will delve into a holistic approach for managing risk within the plant and throughout the whole organization. Monitoring feed stocks, in process parameters, and maintenance data using analytical tests creates an overall view of plant process stability. We will explore case studies and best practices for managing and guiding plant decisions creating a more safe environment for both assets and people.

480108 Data-Driven Root Cause Analysis for Event Investigation Followed By Real-Time Monitoring and Prevention (extended version)

Abstract:

Modern chemical process units have several thousand sensors installed that stream data into the Distributed Control Systems (DCS). The sensor data is then accumulated in historian databases spanning multiple years. Aspen Technology has recently introduced powerful new software that allows for causality analysis based on data patterns that appear when unique process events occur. Without the loss of generality, we focus on negative events (process upsets, shutdowns or equipment malfunctions), but the alternatives can be studied as well, e.g. an exceptionally good batch, or a faster than average plant startup. The starting point of investigation is a dataset that consists of a sufficiently long segment of continuous historical data record with good temporal resolution (e.g. minute data for one year or more). We represent the dataset as a collection of synchronized time series, for the available historian tags, each representing a process sensor. While it is not the only scenario, a negative event is represented as a particular pattern within a single tag. Such a tag might track some KPI parameter such as product purity or power consumption, etc. A fast parallel algorithm analyzes all tags within the dataset. The number of tags might be in thousands. The result of highly parallel analysis is a model that contains causal precursor patterns to precede the negative event. Our algorithm provides a natural view on sensors (tags) and their dynamics and thus leads to a streamlined analysis of precursors from a user perspective. Post training, our root cause analysis model is available for the real-time deployment. In real-time deployment, given an observation of any combination of precursors, we compute the probability of event to occur at some time in future up to a maximum time horizon. We can then observe the propagation of high probabilities of the event at some distant future (at time horizon) towards present time. Various warnings based on probability levels at different time spans (towards present time) can be created to monitor and prevent negative events.

480668 Moving the Needle with All Data, Big and Small

Abstract:

Dow combines the power of science and technology to passionately innovate what is essential to human progress. As a result, data science has been a vibrant theme at Dow for decades, in both the R&D and manufacturing organizations. More recently, the Advanced Analytics team was formed to apply these skill sets to answer broader business questions. Join us in this session to learn about Dow's analytics journey, using big and small data technology to drive growth. Through forecasting, data and text mining, and optimization and simulation methods, the Dow Advanced Analytics team is poised to add tremendous value to the company's top line, while becoming an essential partner for decision makers across every level of Dow.

478986 Digital Transformation in Pharmaceutical Product Development - Value, Strategy and Implementation

Abstract:

Pharmaceutical product development is the act of selecting and transforming a molecule to a valued medicine subject to patient, regulatory, manufacturing and business requirements. Organic Chemists, Molecular Biologists, Chemical Engineers, Mechanical Engineers and Analytical Chemists work together to develop the product design and its critical quality attributes, the process design and its various critical process parameters to deliver the right medicine in the right presentation with the right product & process knowledge to the right patient at the right time. The digitization of the underlying data is a monumental task, and requires a carefully conceived strategy. This talk will articulate the business drivers for this digital transformation, the three key elements of the strategy, namely Data Integration, Data Structure, and Analytics, and the progress on the implementation plan within Lilly.

480533 Big Data Analytics for Batch Performance Improvement

Abstract:

Batch processes are ubiquitous in food, chemical, pharmaceutical and semiconductor industries. Big Data Analytics efforts in chemical industry in particular have largely focused on continuous processes up to this point in literature (1–3). In this paper, we present some of our findings in using Big Data Analytics methods to explore batch-based datasets. We present two case studies in using batch datasets to improve batch production by identifying key steps and variables associated with longer cycle time.

In the first case study, we examine a batch system with multiple operating units. The general approach consists of: (1) identify general step sequences, (2) assess the correlation between steps and other common process interruptions, and (3) identify key variables in the steps of interest for improvement. The methodology is helpful to understand the sources of the highest variability that affect the batch quality variable of interest. The proposed methodology is validated on an industrial system where three batch units (with over 100 variables and many recipe steps) are operating in series. After performing the data analysis and identifying critical steps and process variables, the production rate of this batch system has shown consistent improvement.

In the second case study, we combine data-driven models constructed from batch process data with a controller to reduce process variation of a batch reaction system. After analyzing the batch dataset from this process, key recipe step was identified to be the source of variability. However, conventional control strategies cannot be applied easily due to the unique process equipment. As a result, we propose an alternative control strategy based on the model from the batch data analysis, where both normal operating data and design of experiment data were utilized. In addition to minimizing variability, the safety and equipment constraints are also considered in the control action, leading to robust and stable control. The effectiveness of the proposed batch control strategy was demonstrated to show considerable improvement through both simulated case study and plant implementation.

References

1. Qin SJ. 2014. Process data analytics in the era of big data. *AIChE J.* 60(9):3092–3100

2. Colegrove LF, Seasholtz MB, Khare C. 2016. Big data: getting started on the journey. *Chem. Eng. Prog.*, March, pp. 41–45

3. Kadlec P, Gabrys B, Strandt S. 2009. Data-driven soft sensors in the process industry. *Comput. Chem. Eng.* 33(4):795–814

480037 The Canopy Platform for Process Monitoring and Optimization

Abstract:

Enthought provides product and consulting solutions for the chemical, petrochemical, pharmaceutical, and semiconductor industries. At the core of these solutions is the Canopy Platform, which brings together a host of powerful components for data analysis, visualization, local and remote data access, and data exploration. Canopy Platform solutions allow engineers to easily browse and visualize huge amounts of data stored remotely, do interactive MATLAB-style analysis on it, and develop analysis or detection algorithms and deploy them at scale over large datasets using cluster or cloud compute resources. Detection and prediction algorithms developed in this fashion can be applied to monitor live incoming data and provide feedback to operators, used to provide input to maintenance scheduling, or used to debug production issues. These solutions place a strong emphasis on both convenient desktop user interface (useful for experts to develop their algorithms) and web dashboards for monitoring or data consumption by a broader audience. Web service components makes detection or prediction algorithms written on the platform accessible programmatically, so that the development effort can be reused by other systems and software throughout the business.

In this talk, we show how we leverage the Canopy Platform to build modern, customized software for analyzing plant process systems. We'll demonstrate the *ad hoc* analysis capabilities and workflow-oriented user interface that simultaneously simplifies open-ended and routine tasks. These capabilities make use of the extensive library of analysis routines that are available to the Canopy Platform and the data management component that enables uniform access to raw files, databases, and remote data stores. We'll walk through a cluster analysis workflow that can make use of a company's existing compute resources to scale to thousands of controllers and years worth of batch or continuous data. We will also describe the larger vision for this software as an automated plant advisor that monitors the status of processes, warns when parameters are out of expected bounds, forecasts failures, and offers solutions to those problems.

480901 Big Data Process Modelling with Parallel Graphics

Abstract:

The scale of data collected across a typical process plant, thousands of points at sub-minute frequencies, for decades, completely swamps engineers' ability to use conventional analysis and graphs. Among the solutions considered are big data approaches that have produced results in other fields. But process plants present big challenges for big data. Big data techniques are focused on detecting small correlations among a mass of largely random data, while process data has extensive correlations due to mass balance and physical relationships. Predictive analytic techniques based on big data provide generalized answers through simplification, ignoring data by choosing a subset of variables, averaging data, and ignoring fundamental complexities and nonlinearity. This destroys the richness of the data and reinforces preconceptions.

Process engineers are skilled at data reduction. When investigating a problem a typical approach is to place a boundary as small as possible around the suspected equipment, select a few variables known to be key and a few more that might be related to the problem at hand, limiting the approach to 10 to 20 variables of the hundreds that may be available and then focusing in on just some key time periods believed to be key. In this way, only a very small amount of the applicable data is ever used and very little new understanding can be generated. The drawback has been that multi-way interactions have exponential complexity with the number of variables and can take prohibitively long to investigate, understand, and discover new relations that aren't just the known and expected physical correlations.

The parallel coordinate graph removes this data analysis limitation. By providing a graph that allows viewing continuous data across hundreds of variables simultaneously, engineering analysis and visualization can proceed roughly linearly with the number of variables considered, dramatically increasing the amount of data available to the engineer and brining the limit closer to the actual physical memory limit of the computer system used. Parallel graphs with queries can be used to link and compare operating envelopes from final product quality variables across hundreds of process operating values, allowing discovery and utilization of data not previously considered.

Extending this approach to geometric modelling, over one hundred variables can be used in a single process operating model built from the same historical data. The use of the parallel coordinate plot allows operators to monitor the process and detect changes in the relationships between the variables in real time. Unlike traditional models which are practically limited to 10-20 variables, this model takes account of far more historical data and variable relationships.

As an example an ethylene refrigeration system will be considered. Here one of the operation issues is keeping the system far enough from compressor surge that the anti-surge system doesn't operate. By including many more variables, a much more sensitive event prediction models can be built that allows more optimal operation while still giving warning of required operator action.

Value of Process Knowledge and Models in Data Analytics Applications

Jaleel Valappil and David Messersmith

Bechtel Oil, Gas and Chemicals

Andri Rizhakov and Kelly Knight

Bechtel Nuclear, Security and Environmental

AIChE Spring Meeting 2017, San Antonio, TX

Session: Big Data Analytics – Industry Perspective

1.0 Introduction

A typical chemical process facility requires large capital spending to achieve full production. To realize maximum commercial benefit from these capital investments over the lifecycle of the facility, it is important to utilize these existing assets to their full potential. Process facilities have relied on traditional DCS/APC systems and enterprise performance management platforms to maximize their asset utilization. There is a great potential to further enhance the value from the existing assets using new industrial internet and big data analytics technologies.

The industrial internet is a recent development enabled by the convergence of cloud computing, big data, sensor technologies and cheap computing resources. This technology utilizes very large amounts of real-time data from diverse sources to build model based analytics for various applications. The industrial internet combines the physical plant with the digital information to derive new benefits. This is enabled by the decreasing cost of sensors, computing storage/ infrastructure and innovations in data analytics [1]. Recent developments in big data analytics enables wider deployment of this technology [4] in process industries.

2.0 Main Applications and Benefits of Big Data Analytics in Process Industry

2.1 Process, Equipment and Control Monitoring

Monitoring of plant data for abnormal operating modes is an important application of big data analytics. Abnormal operating scenarios result in lost plant productivity. Identifying these conditions and taking corrective actions early before it results in plant downtime is hence very valuable. One of the key applications of industrial internet and big data analytics is real time monitoring of the performance of the process and equipment using the data collected. The data can be used to build either predictive or unsupervised models to monitor the state of the process and equipment and trigger alarms for operator to take corrective actions.

2.2 Asset Management - Predictive Maintenance

Predictive maintenance is the most important application of industrial internet and data analytics that enable identifying equipment problems beforehand and scheduling maintenance at appropriate time. This involves monitoring equipment conditions continuously and predicting when a failure is likely to occur. Traditional way of performing maintenance is preventive, whereby the maintenance is performed at pre-scheduled intervals. This can result in unneeded maintenance at times. In other times, preventive maintenance may be too late if failures result from events or conditions that occur at any time. Predictive maintenance enables maintenance planning based on actual condition of the equipment. This also avoids the unplanned downtime that can be caused by reactive maintenance. Predictive maintenance is not limited to turbomachinery. Other process equipment, including exchangers and valves can benefit from this. Some of the critical valves can be proactively replaced before failure to minimize unplanned downtime. Heat exchangers can be monitored with additional wireless temperature sensors and cleaned before the fouling starts affecting plant production.

2.3 Plant Operational Optimization

This application of big data analytics involves giving operational guidelines to maximize production, energy efficiency or other economic objectives. Currently, most of the process plants in operation are equipped with Advanced Process Control (APC) systems, which maximize the capacity or efficiency. These systems are layered on top of DCS and execute every minute or so. Apart from this, some of the operating plants are also be equipped with RTO (Real Time Optimization), supervisory or advisory systems that aim to enhance the operation of the plant. There is further opportunity to optimize smaller units in the plant. It needs to be noted that these are currently supervisory functions that provide recommendations to the operator (Unlike APC systems, which write back to the DCS on a real time basis).

3.0 Value of Process Data and Knowledge for Big Data Analytics

To realize the maximum return from investments this emerging technology area, LNG industry needs to properly utilize the existing plant information and resources. There are several valuable resources in operating facilities and plants under construction that can be beneficial in this regard.

1. **Process knowledge and information**

   Process design and operations knowledge is an important asset that can enable realizing the maximum benefit from investment in industrial internet. Unlike other domains, process facilities are inherently constrained by thermodynamic relationships. Understanding underlying process relationships can streamline application of this technology.

   There are two areas where process knowledge is important:

   1. This is a key input for the selection of data analytics application and this helps to establish and select the right variables to be included in the data driven models.
   2. This is also critical in identifying the opportunities for optimization applications and selecting the right objectives and scope of the same.

   Process plants recently completed or under construction collect the relevant design data in digital format that is used for engineering. The Smartplant Foundation or similar platforms are used to store 3D models, datasheets and P&ID information with digital links among them. These provide a virtual information repository that can be leveraged to benefit the development and deployment of big data applications.

2. **Fundamental simulation models**

   Process facilities utilize steady state and dynamic simulation models during EPC phase to enhance engineering design and ensure safety. In addition, majority of the plants develop Operator Training Simulator (OTS) to assist in training and startup. These models and OTS can be beneficial in maximizing the value derived from industrial internet. In addition, industrial internet facilitates the deployment of digital twins, which are exact replica of a key plant asset that can be used for monitoring, optimization and troubleshooting.

   The fundamental process models are valuable in various ways in enhancing the data driven models used for industrial internet applications, including:

   a. Selecting the right model structure: Many of the data driven models (Especially for optimization) need causal models with the correct cause and effect. Identifying this from data can be near impossible in certain cases. Process models can identify the correct causal links and also establish the variables that are significant. This feature selection is an important step in data analytics applications.
   b. Boosting the data models: There are instances where sufficient data that cover the operating range may not be available from operational data. In these case, simulation runs can augment the data to provide the required range.

c. Model Based Optimization: The availability of models and the ability to host these remotely opens up new possibilities to optimize the process remotely. These models can be used for steady state optimization to establish the best operating parameters for the plant.

4.0 Case Study – Amine Plant Monitoring

There are several advantages to using data driven analytics for monitoring industrial systems. Two different approached can be used for fault detection in process plants:

a) Model based – Which uses fundamental process models to detect deviations from normal operating conditions.
b) Data driven – Which uses historical process data for fault detection.

The application of graphed anomaly detection techniques in determining abnormal operating conditions in LNG process is presented below. This is applied to amine plant in a currently operating LNG facility. A predictive model consisting of statistical probability distribution (Bayesian distribution modeling) is developed form historical data. The distribution is not limited to normal and can have different shapes. A normalized anomaly score is calculated. The advantages of using the technique presented here include:

a. It reduces plant downtime by identifying operational problems in advance
b. It helps to identify root cause of problems quicker
c. Can be done without historical data, as it learns online
d. Can be set up easily without large investment

One of the key steps in the graphed anomaly detection is the clustering of variables for anomaly detection. Statistical methods like k-means and k-medoids can be used for clustering. If design data is available, that can be used to set up clusters based on equipment. Alternately, the clusters can also be developed utilizing the variables selected based on process knowledge.

4.1 System components and setup

The system components utilized for anomaly detection is shown in Figure 1.

- Splunk is used to index the machine data coming from the LNG process. Splunk is a widely used big data tool and has the ability to index various types of data including streaming data.

- Prelert is used to detect anomalies using machine learning techniques. Prelert is a statistical anomaly detection technique, which develops probability distribution of the variables and updates it as new data comes in. An anomaly score is determined based on the distribution and the incoming real-time data [2].

- Neo4j is the graph database used to store and analyze the design data from Smartplant Instrumentation database and use it for anomaly detection.

4.2 Application to amine system monitoring

Application of above tools for detecting anomalies in amine system used in LNG plant is studied. Amine system removes CO_2 and H_2S from natural gas before it is send for liquefaction. The performance of this unit is very important as any carried over CO_2 from amine unit can freeze in the cold temperatures in liquefaction section causing plant downtime and lost production.

The historical data from an operating LNG plant was used for this analysis. The data used corresponded to a seven-month period of operation after startup. The relevant tags were identified and corresponding data was collected. In addition, design data related to instrumentation and other items were obtained from Smartplant Instrumentation database to aid this project.

The analysis for amine plant data was initially performed using all the data (About 300 tags) for the unit to determine an anomaly score. This was fund to be unsatisfactory as the score was not capturing changes in the plant operation.

The anomaly scores were also calculated using data clustered by key equipment (Amine absorber, amine regenerator etc.). The tags that belong the individual equipment are derived from Smartplant database. A score is calculated for each equipment in this case.

As a third approach, data clustered by groups as determined from operational experience was used for determining the scores. Four different groups related to acid gas extraction, acid gas removal, separation performance and amine inventory were identified for this purpose.

An example of data driven anomaly detection performance is shown in Figure 2. During the operation, the CO_2 composition was found to increase to above 100 ppm while the plant feed gas rate was close to the normal rate. This indicated that there was some upset condition that may be causing this. The inlet feed gas CO_2 composition was found to increase during this period, which affected V-1201 operation (Acid gas removal column). The results shown in Figure 2 demonstrates that the anomaly score specifically for the equipment V-1201 did show an increase well before the CO_2 concentration (red line) increased. This could have been used to alert the operators about the impending problem.

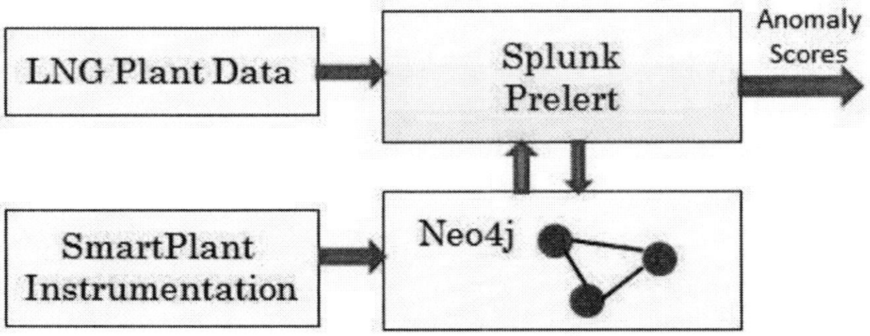

Figure 1: System architecture for anomaly detection

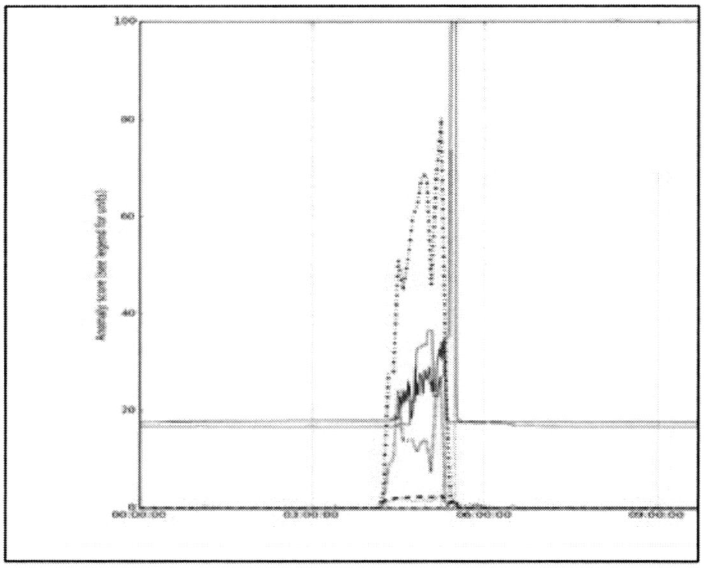

Figure 2: Result of anomaly detection. The dotted blue line shows the score.

5.0 References

1. "*Industrial internet reference architecture*", Published by Industrial Internet Consortium (IIC).
2. Veasy, J. and Dodson, S., "*Anomaly detection in application performance monitoring data*", International Journal of machine learning and computing April 2014, 120-126.
3. Johnson, D., "*IIoT applications enable the process world*", InTech Magazine, Mar-Apr 2016.
4. Perrons, R.K. and Jensen, J.W., *"Data as an asset: What the oil and gas sector can learn from other industries about "Big Data""*, Energy Policy, 81(2015) 117-121.

Connecting the Dots – Gaining Insight from Big Data in Hydrocarbons Manufacturing

Birgit Braun, George Bellos, Stef Luijten, Liwei Li, Mark McAdon, Jason Hamilton

AIChE Spring Meeting 2017

The Dow Chemical Company

"Big Data" and Manufacturing

Big Data = "data sets too large and / or complex for traditional data processing tools and applications" (think Excel)

→ *common and not new in manufacturing*

➢ Large data sets
- Process data, analytical data, meta-data → data bases
- Global business → about 200 furnaces across 14 plants

➢ Complex data sets
- Different routine process scenarios, feeds, technologies
- Expected and unexpected non-routine operations

New Opportunity

New opportunities for manufacturing data

- ➢ Manage different data repositories
- ➢ Combine / align / refine data
- ➢ Modeling and visualization
- ➢ Improve understanding

3

Graphic: https://inspiredittrainer.com/2013/07/28/what-is-big-data/

New Opportunity

New opportunities for manufacturing data

- ➢ Manage different data repositories
- ➢ Combine / align / refine data
- ➢ Modeling and visualization
- ➢ Improve understanding

- **Multivariate relationships**
- **Better forecasting**
- **More reliable (real-time) monitoring**

4

Graphic: https://inspiredittrainer.com/2013/07/28/what-is-big-data/

Example Applications in Hydrocarbons Manufacturing

- Furnace technology and operation
  - Coil technology differentiation
  - Performance over time
  - Improve process conditions
- Process chemistry
  - Key process / multivariate indicator identification
  - More reliable fouling monitoring

Innovative Approaches and Solutions

Internally created platform

- Interfaces automatically with updating databases
- Integrated code for date alignment and pre-processing
- Streamlined data modeling
- Visualization of results

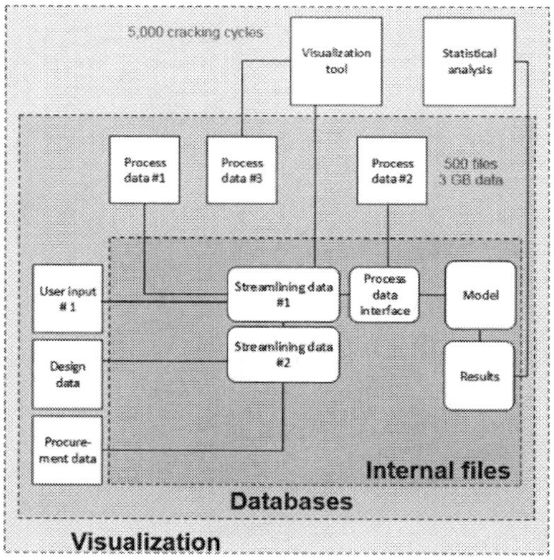

Example: Furnace Runlength vs Different Furnaces

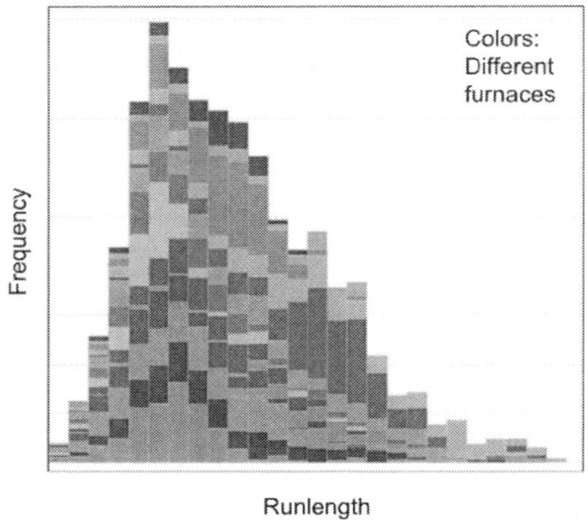

- Multiple years of plant performance data color coded for different furnaces
- Not all plants perform equal
- Identification of biggest improvement opportunities

7

Example: Furnace Runlength vs Different Technologies

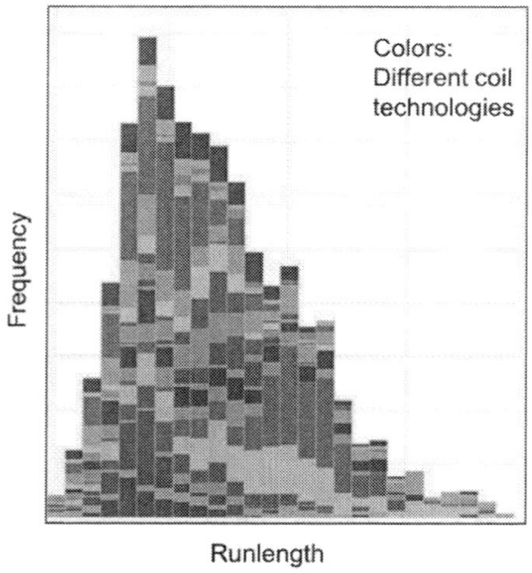

- Multiple years of plant performance data color coded for different coil technologies
- Run length is different for different coil technologies
- Initiation of possible root cause investigation and improvement initiatives

8

Example: Technology Impact on Coil Age

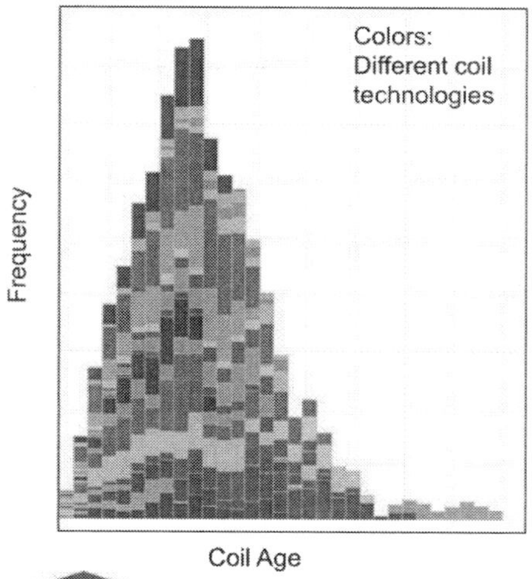

- Multiple years of plant performance data color coded for coil technology
- Shows differences in typical coil age for various technologies
- Identification of advantaged coil technologies

9

Optimize Practices

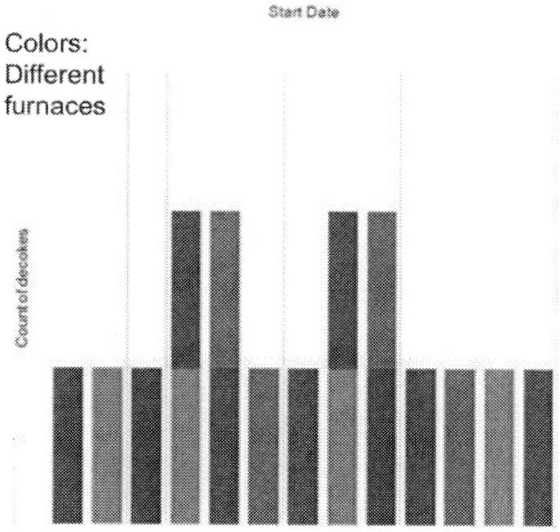

- Identifying furnaces going through simultaneous decoke cycles
- Challenge operating practices

10

Example: Variation of Process Conditions per Plant

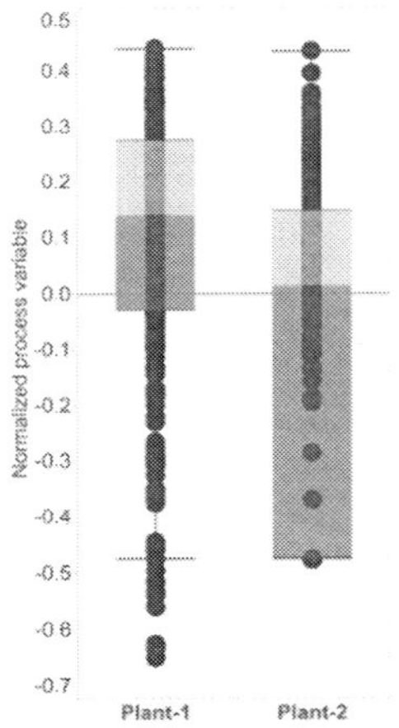

- ➢ Identification of different operating practices
- ➢ Identify units running outside recommended range
- ➢ Contribute to reliability improvement

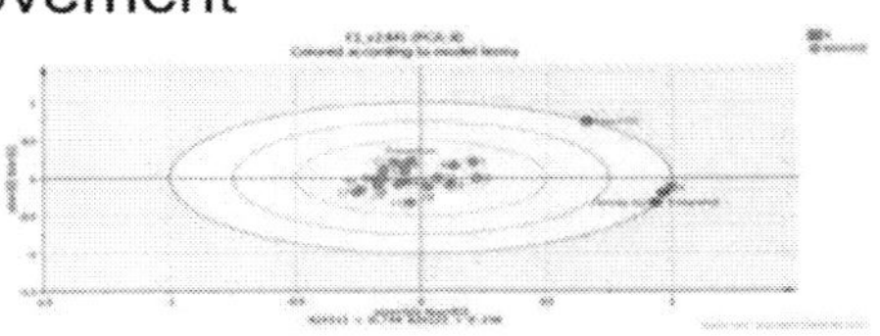

11

Information at Your Fingertips → Interactive Dashboard

Dashboard

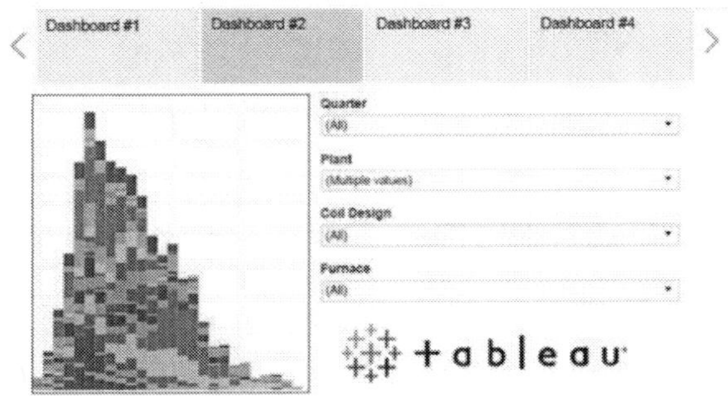

- ➢ Share information with a group
- ➢ Tailor made data selection
- ➢ Access to 5Gb+ data in few seconds

12

Identification of Differentiating Patterns

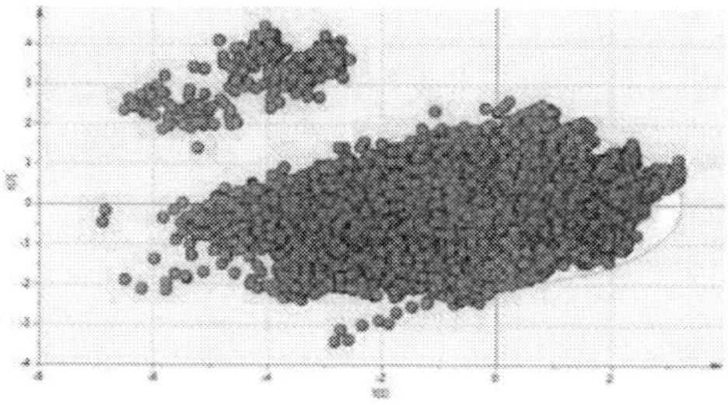

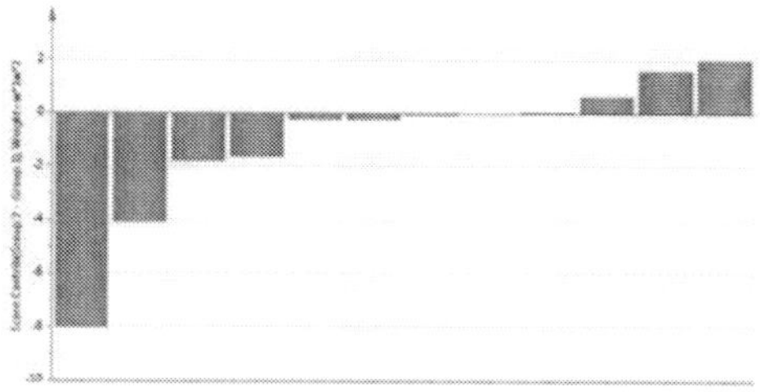

- Pre-processed data sets for efficient PCA analysis
- Identification of different operating conditions and technologies on plant performance

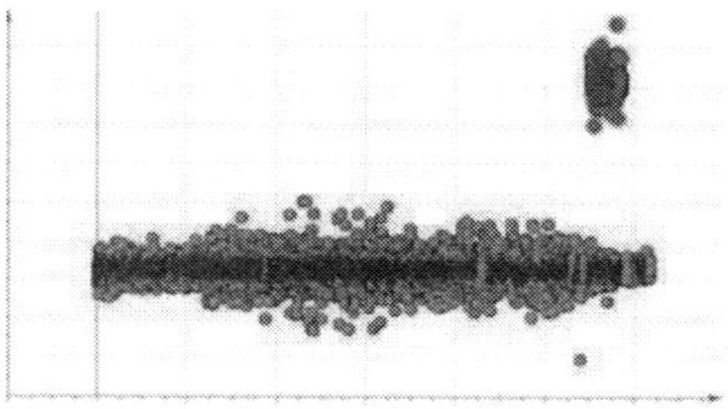

13

Information Diluted in Noise
Following Up Performace of x Radiant Coils

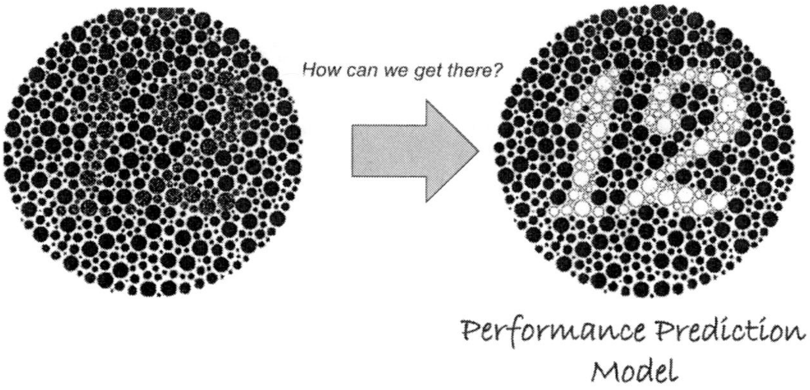

Pictures are taken from https://en.wikipedia.org/wiki/Ishihara_Test#

14

Performance Prediction Model

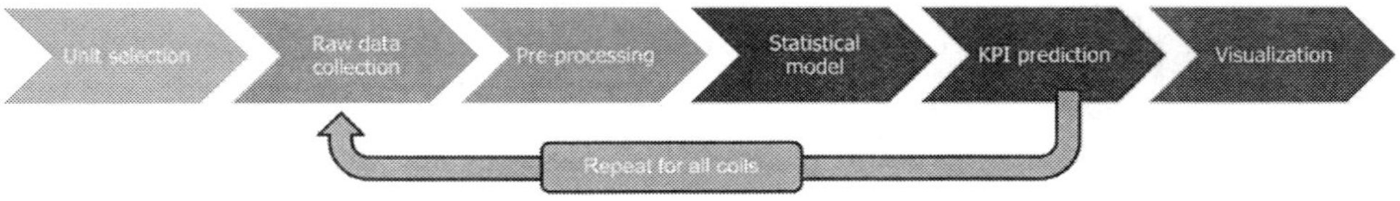

- Unit performance evaluation
  - Different runs
  - Selectable time during run
- Improve process knowledge

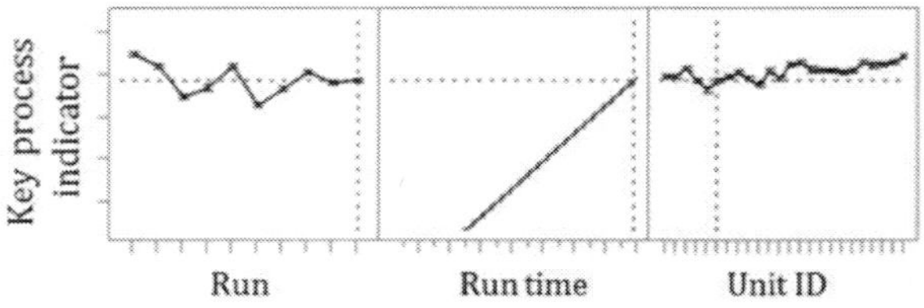

jmp Statistical Discovery.™ From SAS.

15

Identification of Relevant Key Process Indicators

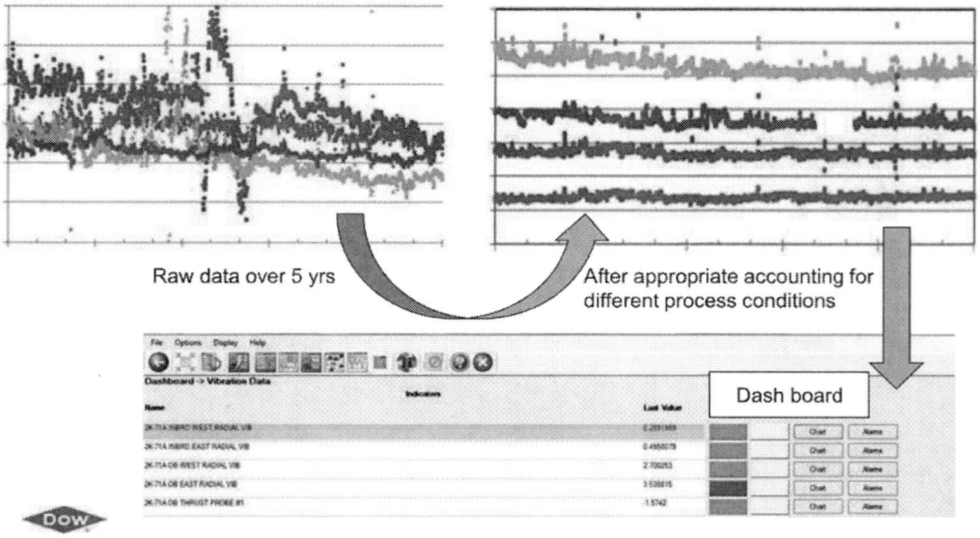

16

Conclusions

- Connecting the dots / data from different data sources enables
  - Technology evaluations
  - More detailed plant performance analysis
  - Operation improvements
  - Better performance monitoring
- Critical for success
  - Team work
  - Appropriate preprocessing
  - Automatic data update

17

Thank You

478954 Real-Time Batch Process Monitoring in Biopharmaceutical Manufacturing: Big Data Analytics for Small Data

Abstract:

In recent years, the biotechnology-based products have been gaining increasing visibility and success in treating chronic diseases such as arthritis, diabetes and cancer. As per the report published in 2005 [1], approximately 27% of the new medicines in active development are in fact biotechnology-based products.

To allow for an efficient monitoring and control of biopharmaceutical processes, multivariate statistical techniques are routinely deployed. Despite over two decades of research and advancements in industrial batch process monitoring and control, biopharmaceutical process monitoring faces a unique challenge due to scarcity of historical data. In fact, it is common in biopharmaceutical manufacturing to have only a few engineering batches for a product with no commercial production history at a manufacturing facility before the actual good manufacturing practice (GMP) campaign. While limiting the number of engineering runs improves speed to market and reduces manufacturing costs, it also limits the amount of representative data available for establishing a monitoring baseline, which presents several problems. First, with small data sets it is nontrivial to capture the inherent variability under the normal operating conditions (NOC) in their entirety. Second, most of the multivariate monitoring methods, such as principal component analysis (PCA) and partial-least squares (PLS) are big-data technologies that are effective only when there are large historical data sets.

In this talk we will discuss several machine-learning methods to deal with the data scarcity problem in biopharmaceutical process monitoring. We propose a non-parametric modeling approach to generate virtual campaign data under normal operations on-demand. Using virtual campaign data, standard batch monitoring techniques are used to effectively monitor a biopharmaceutical manufacturing process. Encouraging results are reported from practice.

References:

[1]. Hamilton G. The Biotechnology Market Outlook: Growth Opportunities and Effective Strategies for Licensing and Collaboration. Dublin: Research and Markets. 2005.

AUTHOR INDEX

AUTHOR INDEX